DIGITAL SAFETY IN THE AGE OF AI: PROTECTING YOUR PRIVACY IN A CONNECTED WORLD

LUCY LIPARI, M.A.

Digital Safety in the Age of AI:

Protecting Your Privacy in a Connected World

Publisher: Buddy Press / American Building Corp.

You should not rely on this book for legal, financial, or professional advice because it is only for general informational purposes. Readers should verify settings and features with their device or service provider because technology changes rapidly.

ISBN (Paperback): 978-1-7362547-8-3

ISBN (Ebook): 978-1-7362547-7-6

ISBN (Hardcover) 978-7362547-9-0

The Digital Safety Series

Stay safe. Stay smart. Protect your digital life.

Digital Safety in the Age of AI

Digital Safety for Teens

To Richard—
For your steadfast support, courage, and unwavering belief in every page I write.
Your strength, humor, and heart are the foundation of everything I create.

And to my readers—
Thank you for your curiosity, open mind, and willingness to explore how technology shapes our lives.
May this book inspire you to protect your privacy, think critically, and never stop learning.

Disclaimer and Author's Note

This book aims to inform and empower, not to alarm or overwhelm. While I've taken care to ensure all examples and recommendations are accurate and up to date, technology—especially artificial intelligence—is growing at an unprecedented pace. The opinions, interpretations, and suggestions in this book reflect the author's own analysis and responsibility.

AI tools and systems are continually being refined and redefined. As a result, some information or examples in this book may change over time as new technologies, laws, and protections emerge.

All insights and explanations are based on the author's independent research, drawing on credible studies, reports, and trusted digital safety sources.

As the author, I am learning alongside you. Each new digital breakthrough or challenge presents an opportunity for us all to become more informed and resilient.

"Digital Safety in the Age of AI" is not a substitute for professional legal, technical, or security advice. The resource is only for informational purposes. If you face issues such as identity theft, data breaches, or cybersecurity risks, please consult a qualified expert or attorney.

My goal is to share practical, research-based knowledge that empowers you to make safer digital choices, regardless of how quickly technology grows.

⚖️ Legal Disclaimer

This book provides information for educational and informational purposes only. It does not constitute legal, financial, or professional advice. Readers should seek professional guidance for matters requiring expert help.
The author and publisher have made reasonable efforts to ensure accuracy and reliability, but they make no representations or warranties regarding completeness, timeliness, or suitability. The guidance offered here reflects the author's experience and research at the time of writing. Check the current laws, guidelines, and software before following any recommendations, as AI, cybersecurity, and privacy are growing rapidly. By reading this book, you agree that the author and publisher shall not be liable for any direct or indirect loss arising from the use or misuse of the information.

All trademarks, product names, and company names mentioned are the property of their respective owners and are used for identification purposes only.

CHAPTER 1

THE NEW CURRENCY: YOUR DATA

In the twenty-first century, your personal data has become the world's most sought-after resource. Every click, every search, every photo you upload tells a story about who you are; what you value, what you fear, and what you might buy next. Businesses no longer need to guess what customers want, because artificial intelligence does that work for them, quietly learning from patterns hidden in your digital life.

This exchange, which has become the foundation of the modern internet, has laid the groundwork for the internet as we know it today. We use free email, social media, streaming platforms, and navigation apps, but in reality, we pay for them with our personal information. Companies call it "personalization." Economists refer to it as the "data economy." In practice, it's a trade: your privacy for access, your habits for service, your attention for profit.

The Invisible Trade

What struck me most while researching this chapter was how often I had traded my own data for convenience without ever consciously deciding to do so.

Each time you open an app or browse a website, information flows in both directions. You receive a service, and a server somewhere in the world receives your data, including your location, age, shopping history, or viewing habits. What makes this exchange invisible is that most people never see where their data goes or how it's used.

Artificial intelligence is central to this exchange. Algorithms analyze patterns across millions of users, including what time you wake up, what you click on first, and how long you pause before scrolling past an image. The system learns what holds your attention. Soon it can predict what it will hold tomorrow.

AI doesn't just process information; it interprets it. It builds a model of you, a digital representation that becomes more accurate each day. That model influences what you see online, what you're offered, and even what you believe about the world.

Your Data Trail

Consider the simple act of checking the weather on your phone. The app uses your location to generate a forecast, but it may also share that data with advertisers, who now know you're in a specific neighborhood. You search for a coffee shop, and AI connects that information to your recent searches for pastries. Suddenly, you receive a coupon for a bakery nearby.

This "data trail" follows you across platforms. Search engines, streaming services, and social networks all gather fragments of your behavior. Each company may only see part of your life, but artificial intelligence stitches those fragments together into a full portrait. That portrait becomes an asset, a product that can be sold, shared, or used to predict your next action.

A music app notices you listen to slow jazz late at night. A wellness platform concludes that you might have insomnia. A retailer then advertises herbal sleep aids. None of these systems talk directly to each other, but AI integrates their insights in milliseconds through data-sharing networks.

The Cost of Convenience

Every innovation that simplifies life also invites new forms of exposure. Smart home devices respond to your voice but also record snippets of your conversations to improve accuracy. Fitness watches track your heartbeat, but that data may one day be analyzed by insurers to calculate risk.

Artificial intelligence makes this possible because it can analyze massive amounts of data more quickly than any human can. The more information it receives, the more accurate its predictions become, and the harder it is for individuals to remain anonymous.

Convenience has become the modern temptation. We trade pieces of our privacy for personalized playlists, instant answers, and predictive text. The trade may seem harmless, but each choice fuels the system that defines us.

When AI Knows You Better Than You Do

In the age of AI, personalization can blur the lines between persuasion and manipulation. The same tools that recommend movies also influence political views, health choices, and even dating preferences.

Consider social media: algorithms track which posts you linger on, what prompts you to comment, and which reactions you express. Over time, AI refines your feed to mirror your interests, or what it assumes your interests to be. The more you engage, the more it learns, and the tighter the feedback loop becomes.

Researchers have found that AI can sometimes predict personality traits more accurately than friends or family. By analyzing patterns in your posts, language, and likes, it can infer your mood, beliefs, and emotional triggers. That insight becomes valuable to advertisers and political campaigns alike.

Imagine an AI system that detects when you're tired after midnight browsing. It might promote comfort-food delivery or entertainment videos rather than educational content, given what you're most likely to choose. It's not evil; it's efficient. But efficiency can reshape free will when invisible cues guide every digital decision.

The Hidden Economy of Data

Data is not just collected; it's monetized. Entire industries exist to analyze, package, and resell information about human behavior. The more precise the profile, the higher its market value.

Artificial intelligence turbocharges this economy by making predictions that drive profit. A clothing retailer no longer needs to guess fashion trends; AI reads millions of social media posts and images to forecast next season's demand. An insurance company can assess risk based on patterns in your online activity. A streaming platform predicts which new series will go viral before it's released.

The currency of the modern world isn't cash; it's insight. AI transforms raw data into predictions, which in turn drive revenue.

However, as data flows through layers of companies, applications, and analytics tools, it becomes nearly impossible to track. Even if one company promises not to sell your information, its partners or advertisers may have different policies. That's why privacy laws and ethical AI guidelines are emerging worldwide: to bring transparency back to a system designed for opacity.

Your Digital Identity

Every time you sign into a service, comment on a post, or upload a photo, you reinforce your digital identity, the version of you that exists in digital form. It's a mosaic made of bits of truth: your shopping habits, your reading preferences, your health interests, your travel patterns.

Artificial intelligence doesn't just record this identity; it interprets it. It assumes, for instance, that someone who frequently searches for travel destinations might be open to luxury credit card offers. Someone who reads about job stress might see ads for meditation apps. AI learns correlations, not context; it sees patterns, not people.

Over time, this digital mirror can shape opportunities. Employers might use AI-driven résumé filters that favor specific keywords. Loan algorithms might use browsing data to assess creditworthiness. The danger lies not in the technology itself, but in the assumptions it conceals.

Why Awareness Matters

Understanding how data fuels AI is the first step toward control. Awareness allows you to make informed choices: when to share information, when to restrict permissions, and when to question what's being offered.

You don't need to reject technology to protect privacy. Small actions, such as adjusting settings, using private browsers, and disabling location when unnecessary, can reduce the amount of data AI learns about you. Each step reclaims a bit of autonomy.

The goal isn't paranoia. It's a balance: to benefit from AI's intelligence without surrendering your individuality.

The New Reality: How AI Shapes Your Digital Life

Artificial intelligence has transitioned from a science fiction concept to a daily reality. It drives your car's navigation, answers your customer service questions, edits your photos, and predicts which products you'll enjoy. But beneath the convenience lies a transformation in how we experience reality itself.

AI as the Digital Filter

AI doesn't just gather information; it curates your world. When you open a news app, an algorithm selects which stories to show first. When you shop online, it decides which items to highlight. The more you interact, the more it learns, shaping your digital environment to fit your interests.

This personalization creates comfort, but also a sense of narrowness. Over time, AI builds what experts call a "filter bubble," a digital echo chamber that amplifies what you already agree with. That means fewer surprises, fewer differing opinions, and a more limited sense of the world.

If a student had previously clicked on environmental activism or industrial innovation, they might see different results when researching climate change. A voter might see one version of political reality, while a neighbor sees another. Both truths feel authentic, but a lens filters them.

Case Study: The Resume Robot

Consider how AI shapes opportunity. Many companies now use automated resume scanners that rank applicants before a human ever reads them. These systems use data from past hires to predict success. However, the AI could replicate that bias if past hiring patterns showed bias.

For example, if a company has historically hired more men in technical roles, an AI trained on that data might unconsciously favor resumes with language or experiences associated with men. The data isn't malicious, but the outcome can be unfair.

Awareness and regulation can correct such bias, but only when users understand that AI decisions are not neutral. They are reflections of human data — imperfect, subjective, and shaped by context.

Case Study: The Health App

AI also plays a role in personal well-being. Fitness and health apps track heart rate, sleep, and diet, then use AI to provide personalized advice. That seems empowering until advertisers or insurers share the data. A person's private health patterns may influence the premiums they're offered or the products they see.

The lesson isn't to abandon these tools but to understand their terms. Knowing who owns your data, how it's stored, and what's shared allows you to use technology on your terms.

Case Study: The Smart Home

Smart homes respond to voice commands, adjust lighting, or order groceries automatically. Behind that ease are AI systems that continuously listen, learn, and adapt. They can distinguish between family members' voices and anticipate routines.

But even convenience has boundaries. Voice recordings, energy usage data, and camera feeds can reveal intimate details of your life. When stored on external servers, they become part of the broader data economy, which is valuable to marketers, hackers, and even law enforcement.

The challenge is to design AI that serves rather than surveils — that remains a tool, not a watcher.

The Human Element

AI reflects what we feed it. It learns from human behavior, inherits our preferences, and sometimes amplifies them. That means each of us plays a role in shaping the digital world through our choices.

When millions of users reward sensational headlines or viral content, algorithms learn to prioritize attention over accuracy. When people click on ads for quick fixes, AI assumes that's what humanity wants. The feedback loop between data and design can either elevate or erode the quality of information online.

By understanding that loop, individuals regain influence. Conscious use, pausing before sharing, and verifying before posting help AI learn better patterns.

Living Wisely with AI

Artificial intelligence is not an enemy; it's a mirror. It reflects the data we create and magnifies our habits. Learning to live wisely with it means using awareness as your most outstanding defense.

You can choose transparency and read privacy policies before clicking "agree." You can limit tracking by using a browser that blocks cookies or a VPN. You can diversify your information sources to step outside algorithmic bubbles.

Education, not fear, is the path forward. The more we understand how AI shapes our digital lives, the more we can shape AI in return.

The Empowered Digital Citizen

In the new reality, data is power, and awareness is freedom. Every person has the right to know how their information is used and to decide what to share with others. By making deliberate digital choices, you become not just a consumer but a participant in shaping an ethical AI future.

The next time you scroll, search, or speak to a smart device, remember that your interaction feeds intelligence — but it also defines it. You are part of the system that teaches machines what humanity values.

CHAPTER 2

HOW WE GOT HERE

For most of human history, privacy was the default. You could read a newspaper without anyone knowing which articles caught your eye. You could walk through a market without leaving a record of every booth you passed. Today, however, the world has flipped: almost everything you do online is recorded, analyzed, and monetized. The invisible marketplace that drives this process is called **surveillance capitalism**, an economy built on capturing human experience and turning it into data for profit.

When the term was first coined by social scientist Shoshana Zuboff, it described a new kind of capitalism in which personal behavior itself became a raw material. With the rise of **artificial intelligence**, the system has become exponentially more powerful. AI doesn't just observe what people do; it predicts what they *will* do, and increasingly, nudges them to do it.

Tech companies realized early that the most valuable resource of the digital age wasn't gold, oil, or even information. It was *attention.* The more time users spent on a platform, the more ads they would see, and the platform could collect more data about what engaged or triggered them. Over time, attention itself became the commodity — a tradable, measurable asset managed by algorithms.

In this new economy, you are not the customer. You are the product.

From Data to Dollars

Every click, search, and scroll adds another grain of data to the mountain that fuels AI. Each tiny action may seem meaningless, but at scale, these fragments form a detailed picture of behavior — your preferences, routines, even your emotions.

Companies don't sell your raw data directly; they sell *access* to you. Advertisers pay for the ability to target people who fit precise behavioral categories, such as "new parents researching sleep remedies" or "young adults interested in electric cars." AI systems organize these categories by processing billions of interactions every day.

Let's say you watch a short video about home gardening. Within seconds, AI tracks your engagement — how long you watched, what you clicked afterward, and what other videos triggered similar reactions. You might then see ads for seeds, patio furniture, or self-watering planters. But the reach goes further: data from one app can be matched with information from others, linking your identity across platforms.

This vast ecosystem is powered by *real-time bidding*, an automated auction that takes place in milliseconds each time you open a webpage or an app. Behind the scenes, dozens of companies compete to display their ad on your screen, bidding based on your predicted likelihood to engage. Artificial intelligence determines your value in that moment, turning human attention into financial currency.

You are the resource being mined, and AI is the miner.

Looking back, it's easy to see the warning signs, but at the time, many of these changes felt incremental, even reasonable.

The Psychology of Attention

Why is attention so valuable? Because it's finite. The human brain can only process so much information at once, and platforms compete for every spare second of your focus.

AI-driven design leverages deep psychological principles to keep users engaged. For example, social media feeds use variable reward systems, the same mechanism as slot machines. Each scroll delivers an unpredictable mix of content: a funny post, a shocking headline, a heartwarming photo. The brain releases a burst of dopamine when something exciting appears, reinforcing the urge to keep scrolling.

This process creates **habit loops.** Open the app → experience a reward → feel a brief sense of pleasure → repeat. Over time, these loops can become automatic, even addictive.

Notifications also play a role. The color red, commonly used for alert icons, triggers a sense of urgency in the brain. AI determines when to send these notifications — not just to inform you, but to bring you back at the perfect moment.

The more time you spend online, the more accurately AI learns what keeps you there. It studies pauses, rewinds, and reactions to craft experiences tailored to your psychological profile. In a sense, it knows which emotional buttons to press and when.

The Role of AI in Manipulation

Artificial intelligence transforms surveillance capitalism from observation into orchestration. It no longer waits to see what you'll do; it actively shapes your behavior to maximize engagement.

The process begins with **prediction models**. By analyzing enormous datasets, AI identifies patterns that signal future actions. If you linger on travel photos, the system might infer that you're planning a vacation. If you read health articles late at night, it might conclude you're stressed or unwell. Each prediction helps refine the algorithm's understanding of you.

But prediction isn't enough — AI needs to confirm its accuracy. To do that, it runs constant **micro-experiments**, adjusting what content you see and measuring your response. This is how A/B testing works: the algorithm learns which version keeps attention longer when two users see slightly different posts or thumbnails.

Over time, AI develops the power not just to predict choices but to *influence* them. It prioritizes emotionally charged content because that type of material — outrage, humor, fear, empathy — tends to generate the strongest reactions. This explains why extreme or sensational posts often rise to the top of feeds.

The result is a subtle but powerful form of manipulation. Users are unknowingly guided toward content that keeps them engaged, even if it causes anxiety, division, or addiction.

The Human Cost

Surveillance capitalism doesn't simply change what we see online; it changes *how we think.*

Constant exposure to algorithmically optimized content reshapes attention spans and emotional patterns. We become accustomed to rapid stimulation, short bursts of novelty, and instant feedback. Long-form reading or deep concentration becomes difficult because the brain adapts to expect constant rewards.

Psychologists call this **attentional fragmentation.** It's not that we can't focus anymore — our focus has been reprogrammed to shift rapidly between stimuli. AI rewards distraction because it generates more measurable actions: clicks, likes, and shares.

Emotionally, this system can foster anxiety, envy, and fatigue. Seeing idealized images or curated success stories creates a distorted sense of normalcy. Meanwhile, the endless scroll prevents mental rest. Every time we engage, we feed the system, and the system learns to feed on us.

But the impact isn't just individual. On a societal scale, the pursuit of attention erodes trust. When algorithms amplify outrage or misinformation to increase engagement, truth itself becomes secondary to virality. The result is polarization, not necessarily because of ideology, but because division is profitable.

Case Study: The Social Media Machine

Consider a social platform used by billions daily. On the surface, it's a tool for connection. Beneath the surface, it's a sophisticated AI laboratory conducting continuous behavioral experiments.

Every post you interact with teaches the algorithm something new. If you like a photo of a friend's vacation, the system notes your interest in travel, leisure, and personal milestones. It then predicts which posts will make you linger, perhaps another vacation photo or a destination ad.

At scale, this generates what experts call **behavioral surplus**. This is excess data collected beyond what's needed to provide the service. That surplus becomes the raw material for AI systems that sell predictions about future behavior.

The irony is that while users think they're scrolling freely, the feed they see is custom-built to keep them hooked. What looks like choice is often choreography.

Case Study: The Shopping Algorithm

E-commerce platforms operate under the same principles. Every search, click, and hesitation informs a vast AI network that calculates what will make you buy.

If you view a pair of running shoes but don't purchase them, AI will follow you across the internet, showing you the same product repeatedly. It knows that familiarity increases trust and that repeated exposure often leads to a sale.

But beyond advertising, AI systems monitor subtle cues — like the time of day you shop or how long you hover over a product. If you tend to buy impulsively at night, the system may time its recommendations accordingly.

These mechanisms create the illusion of personalization while masking the underlying manipulation. The goal isn't to help you find what you need; it's to keep you in the cycle of attention that drives profit.

Case Study: The Streaming Trap

Streaming platforms also rely on surveillance capitalism, though in a subtler form. AI tracks your viewing habits, genre preferences, and even the time you stop watching a show. It uses that data to recommend the next title — not necessarily the best one, but the one most likely to keep you watching.

The auto-play feature removes friction, ensuring you never leave the platform. The next episode begins before you can decide to stop. What feels like relaxation is designed carefully to engage you.

In one internal study, a streaming service found that even a 10-second delay between episodes led to a steep drop in viewing time. So, engineers removed the gap, and total watch hours soared. That minor tweak translated to millions of dollars in additional ad revenue and subscriptions.

Again, AI learns not from your preferences, but from your behavior. It knows when you binge, when you browse, and when you break, and it adjusts accordingly.

Ethical Dilemmas

Surveillance capitalism raises fundamental ethical questions. Should a company have the right to collect intimate behavioral data without explicit consent? Do companies have the right to use AI to manipulate attention purely for profit?

Many platforms justify their practices by claiming users agree to the terms and conditions. Yet, these agreements are often thousands of words long and filled with legal jargon that few people read. The AI economy deliberately obscures understanding, which is necessary for valid consent.

There's also the issue of **algorithmic bias.** AI systems trained on historical data inherit the biases of that data. They may reinforce stereotypes, marginalize minority groups, or privilege certain viewpoints. When these biases operate invisibly, they can shape opportunity and perception on a massive scale.

Finally, there's the question of accountability. When an algorithm causes harm, through misinformation, addiction, or discrimination, who is responsible? The engineers who built it? The company that deployed it? Or the user who clicked "accept"?

These are not theoretical debates; they are the defining moral challenges of our digital era.

The Price of Attention — And How to Reclaim It

Surveillance capitalism thrives on distraction, but awareness can break the cycle.

The first step is recognizing that your attention has value. Each moment of focus is a unit of currency in the digital economy. When you choose where to spend it, you shape the incentives that guide technology.

Start by observing your own patterns. Which apps and websites consume the most time? What emotions do they evoke: calm, curiosity, anxiety, envy? Awareness turns unconscious habits into conscious choices.

You can also take practical steps:

- **Adjust privacy settings** to limit data collection.
- **Disable unnecessary notifications** that interrupt focus.
- **Use browser extensions** that block trackers or display how many companies monitor each site.
- **Set boundaries**, for instance, checking social media only at specific times rather than reflexively.

Beyond personal habits, change must also come from public demand. Supporting policies that promote transparency, ethical AI design, and digital literacy helps shift power back toward users. When society better understands how attention is monetized, companies will find it harder to exploit it.

Most importantly, reclaiming attention is not about rejecting technology but redefining our relationship with it. AI can serve as a tool for creativity, education, and connection — but only if used consciously.

Ultimately, your attention is the thread that weaves meaning into the digital world. Every time you pause before clicking, every time you choose depth over distraction, you reclaim a small piece of freedom.

The surveillance economy may watch you, but it cannot define you — unless you let it.

CHAPTER 3

WHAT COMPANIES KNOW ABOUT YOU

For generations, privacy was an expectation of ordinary life. People could close their doors, make a phone call, or send a letter and trust that their thoughts remained their own. Today, however, privacy has transformed into something fragile and uncertain. Every search, message, and photo exists somewhere in a digital cloud, recorded and retrievable.

We live in an era when convenience often disguises surveillance. Every "agree" button we tap opens another door into our personal world. We believe that privacy settings, encrypted apps, and clever passwords protect us, yet these measures offer only partial defense. The truth is that once information enters the digital space, it becomes part of a vast ecosystem of data collection, analysis, and sharing.

The illusion of privacy persists because modern life depends on connection. Our phones track our movements, our devices recognize our voices, and our social media posts reveal more than we realize. What was once private behavior now creates public data trails that define who we are in the eyes of algorithms.

The convenience of digital life has come at the cost of personal boundaries. And though many people sense that they are being watched, few truly understand how deep the observation goes.

The Age of Exposure

Every device you own tells a story about you. Smartphones record not just phone calls but also steps walked, locations visited, and sleep cycles. Smart televisions note which programs you prefer. Fitness bands measure heart rates and moods. Even thermostats and doorbell cameras collect data about who comes and goes from your home.

This web of interconnection creates what experts call the "Internet of Things," a network of devices that constantly communicate. The result is a digital portrait of your daily routine so detailed that it can predict when you are likely to wake, what you might eat, and where you might travel. Artificial intelligence processes this constant flow of information to make convenient recommendations, but it also turns ordinary life into measurable behavior.

Consider a morning routine. When you wake, your smart assistant greets you and reads the weather. Your phone tracks your movement from bedroom to kitchen, while a streaming app records which podcast you choose. As you drive to work, navigation software sends real-time traffic data to your location provider. Each of these systems functions smoothly because of AI coordination. Yet together, they also form a record of your habits that companies can analyze, sell, or share.

The idea of privacy fades not because of a single invasion but because of accumulation. Tiny pieces of information, when combined, reveal who you are with surprising precision.

How Privacy Became a Commodity

I recall pausing mid-draft to review my own account dashboards and realizing I had never fully examined them before. In the digital economy, privacy has become something that can be traded. Every social media platform, free app, or online store operates within a business model built on information exchange. When you sign up for a service, you are not only using a product; you are contributing to a system that profits from knowing you.

Data brokers purchase and compile enormous datasets about individuals. They combine online activity with public records, shopping receipts, and even loyalty card data. These records are then sold to advertisers, insurance companies, and political organizations seeking to influence behavior.

What makes this process so powerful is artificial intelligence. AI organizes data faster and more accurately than humans ever could. It identifies relationships between behaviors — how one purchase predicts another, or how a change in routine might indicate a life event such as moving or changing jobs.

A simple example illustrates this clearly. A young woman who begins purchasing vitamins, reading parenting blogs, and using a baby-name app might be identified by AI as pregnant long before she shares the news with anyone. Advertisers receive that prediction and begin promoting maternity products. In some famous cases, families learned of a pregnancy only after targeted ads appeared in the mail.

This is the new face of commerce: predictive marketing powered by AI insights derived from personal data. Privacy is no longer a right we assume; it is a resource others seek to exploit.

AI and the End of Anonymity

Artificial intelligence has changed the nature of anonymity. In the past, a person could blend into a crowd or browse online without revealing their identity. Today, AI-driven systems can recognize faces, voices, and even walking patterns.

Facial recognition technology, once limited to law enforcement or border control, is now built into everyday devices. Phones unlock by scanning a user's face, while security cameras in stores or airports identify known individuals. These tools provide convenience and safety, but they also create permanent records of movement and appearance.

Voice recognition systems, such as digital assistants, analyze tone, accent, and inflection. They can differentiate between family members, creating personalized profiles for each person in a household. Meanwhile, predictive analytics can identify individuals based on browsing habits or location histories, even when names are removed.

In other words, true anonymity is disappearing. A person can delete an account or clear a browser history, but AI can still recognize behavioral fingerprints — the unique combination of choices, searches, and times of activity that reveal identity.

This is particularly concerning when governments or corporations use AI for surveillance. Predictive policing systems, for example, analyze social and environmental data to anticipate crime, but they also risk reinforcing bias and tracking innocent individuals. Similarly, large-scale data collection during pandemics or political unrest may begin with public safety goals but expand into permanent monitoring.

The line between protection and intrusion is becoming increasingly blurred, and the loss of anonymity alters how people behave. When individuals know they are being observed, they act differently. Creativity, dissent, and free expression can fade under the weight of constant visibility.

The Emotional Cost of Surveillance

The erosion of privacy not only affects technology but also alters human psychology.

Many people feel watched, even when alone. They censor their posts, moderate their opinions, or hesitate before searching specific topics online. Awareness of observation leads to self-control, a phenomenon known as the surveillance effect.

In workplaces, employees monitored by keystroke trackers or camera systems report higher stress and lower job satisfaction. Students using remote learning platforms often express discomfort with eye-tracking features that measure attention. The sense of being evaluated can transform everyday activities into performances.

Privacy is more than secrecy; it is a form of personal freedom. It allows people to experiment, learn, and make mistakes without fear of judgment. When that space disappears, authenticity suffers. The internet once promised expression without limits. Instead, it has created a stage where everyone performs for invisible audiences of data collectors.

The emotional toll is cumulative. Over time, constant exposure leads to anxiety, burnout, and distrust. People may see their relationships change when they unintentionally reveal or misuse personal information. Digital life becomes a tradeoff between visibility and vulnerability.

Case Studies: Privacy in Everyday Life

Social Media:
A young professional shares photos of family gatherings online. Without realizing it, the location metadata embedded in the images reveals their home address. AI systems used by advertisers detect patterns—such as furniture types, neighborhood characteristics, and

even approximate income levels—to target high-value marketing. What began as a moment of joy becomes part of a commercial dataset.

Workplace Monitoring:
In corporate environments, AI-powered software tracks employee productivity. Programs also track how long employees leave emails unread or how quickly they complete projects. Though framed as efficiency tools, these systems can create pressure to perform continuously. Employees begin to prioritize measurable output over creativity or collaboration. Privacy in the workplace fades as personal habits become data points.

Smart Home:

Connected devices promise comfort and control. Yet a smart speaker that listens for voice commands also hears conversations. A refrigerator that tracks expiration dates may share usage data with manufacturers. In one documented case, law enforcement requested data from a home assistant during a criminal investigation. These examples highlight a broader issue: when private spaces become digitally integrated, they are no longer fully private.

Each case shows a consistent theme. The boundary between public and private has become porous, and AI accelerates this erosion by turning routine actions into measurable data.

The Global Privacy Divide

Worldwide, countries do not protect privacy equally. Different nations approach digital rights through cultural and political lenses, creating what experts call the global privacy divide.

In the European Union, strict laws, such as the General Data Protection Regulation (GDPR), afford individuals significant control over their personal data. People can request to see, correct, or delete the information companies hold about them. The regulation emphasizes transparency and imposes severe penalties for violations.

In contrast, the United States relies heavily on corporate responsibility rather than comprehensive federal law. Data protection varies by state and industry, leaving sizeable gaps in oversight. U.S.-based technology companies often shape global privacy standards, prioritizing innovation and profit over restrictions.

Other regions face distinct challenges. In China, large-scale government surveillance integrates AI, facial recognition, and location tracking to maintain social control. In developing nations, digital infrastructure grows faster than regulation, creating opportunities for exploitation.

These differences highlight a troubling truth: privacy has become a privilege. The ability to control personal information often depends on geography, income, and access to education. Those with the fewest resources frequently have the least protection.

As AI technologies expand globally, the need for universal privacy standards becomes more urgent.

Reclaiming Privacy in a Transparent Age

Although the erosion of privacy can seem inevitable, individuals are not powerless. Awareness remains the foundation of digital self-defense. Understanding how and why data is collected enables individuals to make informed decisions about what to share.

Start by examining digital habits. Review which apps have access to your camera, microphone, or location. Revoke permissions that are unnecessary. Adjust privacy settings to limit data collection, and consider using a browser that blocks third-party trackers.

Cultivate digital mindfulness. Before posting, ask whether the information could reveal more than intended. Insignificant details, such as location tags, birthdays, or travel schedules, can serve as valuable clues for companies and criminals alike.

Technology can also protect privacy when used intentionally. Encryption tools, secure messaging apps, and password managers help preserve control. Using a virtual private network can disguise location data, reducing exposure to tracking.

However, lasting change will require more than personal vigilance. Governments, educators, and industry leaders must promote digital literacy and ethical standards for AI. Schools can teach students about consent in the digital environment, while policymakers can enforce transparency in data-collection practices.

Ultimately, reclaiming privacy is about restoring trust; trust in technology, in institutions, and in ourselves. Privacy allows for introspection and authenticity, which are essential to human dignity.

In this transparent age, connection does not have to mean exposure. With awareness and collective effort, people can create a digital world where sharing is a choice rather than a condition, and where privacy remains a right, not a memory.

CHAPTER 4

AI, ALGORITHMS, AND INVISIBLE DECISIONS

This chapter forced me to slow down and think carefully about how much influence we've handed over to systems we rarely question. In just a few decades, artificial intelligence has moved from science fiction to daily reality. What once belonged to futuristic novels and laboratory experiments now powers the devices we hold, the systems that guide our cities, and the decisions that shape our world.

From voice assistants that answer questions to algorithms that recommend music, diagnose illnesses, or drive cars, AI has become an integral part of modern life.

We live surrounded by invisible intelligence. It routes our phone calls, filters our emails, predicts what we will type, and manages the traffic we drive through. In many ways, it has become a silent partner, helping us work more efficiently, learn more effectively, and stay connected.

Yet behind this remarkable progress lies a growing unease. The same technology that heals, teaches, and entertains can also deceive, manipulate, and divide.

AI is a double-edged sword: it can amplify human potential or magnify human flaws. Understanding that balance is one of the defining challenges of our time.

The Promise of Progress

Artificial intelligence represents one of the most significant engines of innovation in human history. It is already reshaping industries and solving problems once thought unsolvable. With great precision, AI reads medical scans, spotting tumors invisible to the eye. It also predicts weather patterns and boosts crop yields in agriculture. In education, it personalizes learning for students who struggle.

AI's ability to learn from data gives it a near-limitless range. It can analyze billions of medical records in seconds, translating that information into recommendations for doctors. It can process language to translate across cultures, helping people communicate who never could before. It can even detect natural disasters early, saving lives by forecasting floods or wildfires.

For individuals, AI brings convenience and access. Smart homes adjust to personal routines. Apps translate languages in real time. Navigation systems map efficient routes while conserving fuel. AI makes life easier, smoother, and sometimes safer.

Many experts view this as the beginning of a new human-machine partnership, one that enables humanity to focus on creativity, empathy, and big-picture problem-solving while machines handle repetitive or data-intensive tasks. The promise is real, and so are the benefits. But progress often comes with consequences that are not immediately visible.

The Hidden Risks of Automation

The same qualities that make AI powerful also make it dangerous when misused or misunderstood. Every algorithm learns from data, and this data reflects the biases, limitations, and inequalities inherent in human society. When the data guides decisions about loans, job interviews, or news stories, it can reinforce bias on a massive scale.

AI also operates at a speed and complexity far beyond human oversight. A single error in a codebase or a flaw in a dataset can cause automated systems to make decisions that no one fully understands. For example, an AI system designed to evaluate job applicants might learn to favor specific keywords that appear more often in male résumés, perpetuating gender inequality without malicious intent.

Another risk lies in the automation of misinformation.

Deep fake videos and AI-generated voices can create convincing but false content, blurring the line between truth and fabrication. In politics, such tools could alter public opinion or damage reputations in seconds.

Economic disruption is another concern. As AI automates routine work, millions of jobs may change or disappear entirely. While new roles will emerge, the transition will be painful for those unprepared.

History shows that technology constantly reshapes labor, but artificial intelligence does so at unprecedented speed and scale. It does not simply replace workers. It replaces judgment, pattern recognition, and creativity once thought to be uniquely human.

The challenge, then, is not whether AI will transform society, but whether society can adapt fast enough to guide that transformation responsibly.

The Human Dilemma

Artificial intelligence mirrors humanity. It learns from us, imitates our choices, and reflects our collective values, both noble and flawed. The human dilemma of AI lies in the tension between empowerment and dependence.

AI systems relieve us of effort, but they can also erode skills. Navigation apps make travel easier, but people forget how to read maps. Predictive text saves time, but it reduces the need to think through sentences. As machines anticipate more of our needs, human intuition and patience may become less effective.

There is also the emotional dimension. Many people form attachments to AI companions or digital assistants. These relationships, though convenient, can blur the line between genuine empathy and programmed response. When an AI "listens," it is not because it cares; it is because its function requires it. Yet users often attribute emotion to algorithms, trusting them as confidants or advisers.

In workplaces, AI now evaluates performance and productivity. Employees are aware that algorithms assess their performance, fostering subtle anxiety about being replaced or reduced to metrics. Creativity thrives in freedom, not under constant measurement.

The dilemma becomes clear: AI can extend human capability, but it can also confine it within systems optimized for efficiency rather than humanity. How we navigate that balance will determine whether technology liberates or limits us.

Case Studies: When AI Helps and When It Hurts

Healthcare Innovation

In hospitals across the world, AI has become a diagnostic partner. A system trained on millions of medical images can detect lung cancer earlier than a radiologist. It can identify genetic disorders through facial recognition and assist surgeons with precision robotics. Patients live longer because machines help doctors see more clearly.

Yet the same technology can expose privacy risks. Sensitive medical data, once shared with third-party developers, may be vulnerable to unauthorized use and misuse. A single breach could reveal personal health histories, insurance details, and genetic profiles. What begins as a medical breakthrough can quickly evolve into an ethical dilemma regarding consent and control.

Education and AI Tutors

AI-powered tutoring platforms adapt lessons to individual students, helping those who fall behind. They can identify weak areas, adjust the pace, and provide encouragement. For children with learning differences, this personalization can be transformative.

However, these systems collect extensive data on student behavior—how long a learner pauses before answering, what mistakes are made, even emotional cues inferred from webcam analysis. That information, if shared or analyzed carelessly, could become a lifelong educational profile accessible to institutions or employers.

Criminal Justice and Predictive Policing

Law enforcement agencies use AI to analyze crime data and predict where offenses might occur. In theory, this helps deploy resources more effectively and prevent harm. Predictive systems often reproduce the same biases present in historical policing. Neighborhoods with a higher number of reported incidents receive more patrols, which in turn lead to more arrests, reinforcing a cycle of inequality.

AI's promise of objectivity can disguise structural prejudice. The lesson is simple: intelligence without ethics can become injustice.

The Ethics of Intelligence

Artificial intelligence challenges long-standing ideas about responsibility. When a machine learns from human data and makes a harmful decision, who is accountable? The programmer, the company, or the algorithm itself?

Ethics in AI must begin with transparency. People have the right to understand how automated systems make decisions that affect their lives. This means companies should explain how algorithms function, what data they use, and what limitations they have.

Fairness is another cornerstone. AI must be tested for bias just as a medicine is tested for safety. Engineers can design feedback systems that detect discrimination and correct it. Governments can create oversight boards to ensure ethical standards are met.

Privacy remains equally important. AI feeds on information, but individuals deserve control over how their data is used. Regulations like the European Union's General Data Protection Regulation have started to set global examples, demanding consent and accountability.

Finally, there is the moral dimension. AI has no conscience. It cannot feel guilt, compassion, or responsibility. It acts according to its programming and training data. Human oversight is therefore essential. Delegating moral decisions to machines—whether in warfare, hiring, or healthcare—risks removing the empathy that defines ethical judgment.

AI should remain a tool for moral humans, not a substitute for morality itself.

Living Wisely with Intelligent Machines

Artificial intelligence is here to stay. It will continue to shape economies, art, medicine, and education. The question is not how to stop it, but how to live wisely with it.

The first step is awareness.
Recognize that every AI-powered convenience, from targeted shopping ads to voice assistants, comes with trade-offs. Each interaction generates data that further trains the system. Knowing this allows users to make intentional choices rather than making them

passively.

The second step is participation.
Citizens, educators, and consumers must be part of conversations about AI ethics. Engineers or corporations cannot be the only ones to manage technology. It belongs to all users, beneficiaries, and those affected by it.

The third step is balance.
Use AI where it enhances life, saves time, supports creativity, or improves safety, but limit its reach when it replaces essential human contact or independent thought. Reading a map, writing a letter by hand, or making an unassisted decision may seem small, yet these acts preserve the uniquely human capacity for reasoning and imagination.

Education is key to living wisely with AI.
Children growing up surrounded by intelligent machines must learn not only how to use them but also how to question them. Critical thinking, empathy, and ethics are the skills that will distinguish human intelligence from artificial intelligence in the decades to come.

Governments and industries also share responsibility.
They need to ensure transparency in AI use, implement safeguards against abuse, and design technologies that respect human dignity rather than exploit it. Innovation without ethics risks turning progress into peril.

Ultimately, AI reflects what we teach it. If we train it with compassion, fairness, and humility, it will amplify those qualities. If we train it with greed, bias, and indifference, it will magnify those instead. The mirror of intelligence is only as honest as the minds that shape it.

The future of artificial intelligence will not be determined solely by algorithms, but by the values we embed within them.
Technology is neutral; intention is not. If we choose wisely, AI can remain humanity's most powerful ally rather than its greatest threat.

Ultimately, the goal is not to fear intelligent machines, but to coexist with them responsibly, to ensure that progress and ethics advance in tandem. Living wisely with AI means preserving what makes us human: empathy, creativity, curiosity, and the ability to care.

That balance, once found, will define not only the future of technology but the future of civilization itself.

CHAPTER 5

THE MYTH OF 'NOTHING TO HIDE'

Picture yourself entering a room with walls entirely made of glass. You can move freely, but you are always visible. Over time, you adjust how you walk, what you say, and even what you think. This is the essence of the psychology of being watched. It is a subtle, constant awareness that changes human behavior.

In the digital world, this invisible audience is everywhere. Every email can be stored, every message logged, every click recorded. Cameras monitor public spaces, algorithms track online habits, and social media analyzes moods through posts and reactions. Even when no one is actively observing, the mere possibility of being watched is enough to alter our behavior.

Psychologists have long recognized that observation can alter behavior. This concept, known as the *observer effect,* suggests that people conform more closely to social norms when they believe they are being seen. In the past, that meant acting politely in public. Today, it means curating identities online, performing for followers, and censoring thoughts that might draw judgment or criticism.

We no longer fear only the eyes of neighbors or employers; we are aware of the gaze of machines — the algorithms that analyze, the cameras that record, and the servers that never forget. The result is a new form of self-consciousness, one that shapes not only how we act but how we think and feel about ourselves.

I used to accept the argument that there was "nothing to hide" without much thought, until I began to see how easily ordinary information can be misinterpreted or misused.

The Evolution of Surveillance

Surveillance is not new. Human beings have always observed one another to maintain order and trust. In small villages, gossip was the earliest form of social monitoring. It kept behavior in check through reputation and shame. People behaved because they knew their community was watching.

As societies grew, surveillance became institutional. Governments created identification systems, censuses, and archives. The industrial age brought factories where managers timed every worker's movement. These methods were visible, and you knew when you were being watched.

Digital surveillance, however, operates differently. It is constant, invisible, and often voluntary. We give away information for convenience, such as GPS navigation, loyalty points, and free cloud storage. In return, companies and institutions learn more about us than we realize.

Artificial intelligence magnifies this by turning passive observation into active prediction. Cameras do not merely record faces; they recognize them. Online platforms not only collect posts but also analyze tone and emotion. AI-driven systems now anticipate behavior before it happens — predicting purchases, habits, and even moods.

What began as a means to improve safety and efficiency has grown into a vast ecosystem of data capture that encompasses every aspect of life. It is no longer just about being watched; it is about being interpreted and influenced.

The Brain Under Observation

Being watched has a measurable impact on the human brain. Neuroscientists have found that the mere perception of observation activates the prefrontal cortex, the part of the brain responsible for self-control and social behavior. When we sense surveillance, we become more cautious, more compliant, and less spontaneous.

This can serve positive purposes. Observation can promote honesty and accountability. For example, when people believe they are being filmed, they are less likely to cheat or exhibit aggressive behavior. But constant observation has psychological side effects. It can increase anxiety, reduce creativity, and create what researchers call *cognitive load* — the mental burden of self-monitoring.

In workplaces that use digital tracking or performance analytics, employees often feel pressured to appear productive, even when the tasks they perform require reflection and rest. In classrooms equipped with facial recognition attendance systems, students report a sense of unease, as though every expression is being graded.

Over time, this awareness reshapes the mind. Creativity thrives in privacy because it requires freedom from judgment. When others are watching, people create what is expected rather than what is original. Privacy gives the brain room to experiment; surveillance encourages conformity.

The digital age has blurred the line between public and private thought. Our devices, always within reach, blur the sense of solitude that once helped ideas grow. The result is a generation of people connected to everything yet rarely alone with their own minds.

The Loss of Solitude

Solitude is not loneliness; it is the quiet space where people process their emotions, form opinions, and gain a deeper understanding of themselves. Historically, solitude was a source of wisdom. Writers, philosophers, and inventors retreated into silence to think deeply. Today, that silence is harder to find.

Every buzz, ping, or notification reminds us that someone, or something, might be trying to reach us. Even when we put devices aside, the knowledge that data continues to flow can intrude on peace of mind. Smartwatches monitor heart rates while we sleep, and fitness apps record steps without our command. Privacy no longer depends on action but on constant passive collection.

Psychologists warn that this erosion of solitude affects emotional regulation. Without moments of mental quiet, the brain struggles to recover from stress. The constant hum of surveillance adds another layer: self-surveillance. People learn to watch themselves, editing behavior to fit perceived expectations.

This internalized gaze, often referred to as the digital panopticon, derives its name from an 18th-century prison design where a single guard could observe every inmate. The power of the system came not from constant watching but from the fear of possibly being watched. Today's digital tools recreate the same psychological condition on a global scale.

When solitude disappears, so does the freedom to imagine, to make mistakes, or to simply be. The price of connection can be the loss of inner life.

Case Studies: Digital Anxiety and Modern Life

Social Media Self-Consciousness

A high school student posts a photo online and spends hours checking for likes and comments. What began as a way to share joy becomes a source of anxiety. Each reaction feels like validation or rejection. Over time, this pattern conditions the brain to seek approval through visibility. Psychologists describe it as *social reward conditioning* —a feedback loop in which digital attention replaces genuine connection.

Workplace Monitoring

A customer service representative works remotely under software that tracks keystrokes, screenshots, and idle time. Though intended to ensure productivity, the system produces tension and fatigue. The employee becomes more focused on appearing active than on doing meaningful work. Studies have shown that surveillance in professional settings often lowers morale and reduces trust between workers and their employers.

Smart Cities and Public Spaces

In some public spaces, cameras equipped with AI analyze crowd behavior for safety and traffic control. Citizens report feeling uneasy, unsure of who controls the data or how the data is used, even though the goal is efficiency. This uncertainty fuels what sociologists call *ambient awareness* — a background sense of being observed that influences everyday decisions, from where people walk to what they say in public.

These examples show that the impact of surveillance extends

beyond technical considerations to encompass emotional aspects. Visibility knowledge changes confidence, motivation, and self-worth.

Introducing the New Social Mirror

The digital world has become a mirror that reflects not only who we are but also how others perceive us. Every post, photo, and comment contributes to an ongoing performance of identity. Algorithms analyze these expressions and decide what versions of us appear to the world.

This process gives rise to a phenomenon known as *surveillance-based self-presentation*. People adjust their online personas to fit the expectations of audiences they cannot fully see. On professional networks, they emphasize success; on personal platforms, they project happiness. The result is a constant negotiation between authenticity and acceptability.

This has profound psychological consequences. When identity becomes curated, self-esteem becomes fragile. Individuals increasingly measure worth through metrics—likes, followers, shares—rather than inner satisfaction. The external gaze defines value, and privacy becomes not only protection from others but protection from self-comparison.

Artificial intelligence amplifies this dynamic. Recommendation systems feed users content that reinforces their existing beliefs and preferences. Over time, this creates what psychologists call *echo chambers,* in which individuals encounter only information that aligns with their existing views. In such spaces, identity becomes narrower, and understanding of others diminishes.

Data systems that monetize emotion now control the social mirror formerly held by friends and family. The result is a society that is more connected yet less self-aware, in which the constant reflection of others' attention replaces genuine introspection.

Emotional Cost of Perpetual Visibility

Constant visibility creates emotional fatigue. It heightens vigilance, the sense of needing to stay alert, even in safe environments. Researchers have compared this to mild chronic stress: the body remains on edge, producing cortisol and reducing overall well-being.

Children and teenagers are especially vulnerable. Growing up under digital scrutiny teaches them to perform rather than to simply be. Privacy, once a natural part of adolescence, becomes a luxury. Parents, teachers, and peers all have access to online lives that were once private. This lack of separation can stunt emotional development and delay independence.

Adults experience a different but related burden. The blending of personal and professional personas online forces constant self-monitoring. Many people describe feeling "on display" because they worry that someone could misinterpret or take a comment or photo out of context.

In a world where memory is permanent, mistakes feel irreversible. The pressure to maintain a spotless digital image can lead to anxiety and burnout. Psychologists refer to this as identity fatigue — the exhaustion that results from managing multiple versions of oneself in public view.

The Need for Psychological Privacy

Privacy is not only a legal or technical concept; it is also a psychological one. It provides a boundary that allows people to separate their inner world from their outer world. Without that boundary, the mind struggles to rest and recharge.

Psychological privacy allows individuals to reflect, grieve, imagine, and heal. It allows for the expression of emotions without judgment and the exploration of ideas without fear. If someone invades this mental space, stress and conformity take over.

The constant presence of surveillance chips away at psychological privacy in subtle ways. Even if no one reads a personal journal stored online, the possibility of it being read changes the honesty of the writing. The same is true for creative work, private conversations, and self-expression.

Rebuilding psychological privacy begins with awareness. When individuals recognize the mental cost of exposure, they can draw more precise lines between what they share and what they keep personal. This is not isolation — it is a protection of mental health and individuality.

Finding Freedom in Awareness

Freedom in the digital age does not mean escaping technology. It means learning to live consciously within it. Awareness transforms observation from a source of fear into a tool of empowerment.

The first step is understanding the systems at play. Knowing that algorithms monitor behavior does not have to create paranoia; it can inspire mindful use. Turn off unnecessary permissions, limit app access, and choose platforms that respect user privacy. Each action reinforces control over one's digital footprint.

The second step is reclaiming solitude. Set aside moments without screens, where thoughts belong only to you. Walk, write, or reflect in private spaces. Solitude is not withdrawal; it is renewal. It restores perspective in a world of constant noise.

The third step is authenticity. Resist the pressure to perform online. Share what has meaning for you, not what others expect. By being genuine, you break the cycle of external validation that fuels anxiety.

Communities and governments also play a role. Laws can protect privacy, but cultural change ensures it thrives. Schools can teach digital mindfulness. Workplaces can respect the boundaries between professional duties and personal lives. Families can model trust by valuing presence over constant monitoring.

Ultimately, the experience of being watched is part of modern life. Cameras, data logs, and algorithms will not disappear. But by understanding their effects, we can choose how much power they hold over us.

The psychology of being watched reveals something profound: surveillance shapes behavior, but awareness reshapes power. When people understand how observation shapes their perceptions, they can reclaim autonomy.

People today discover freedom not by hiding, but by knowing, knowing what they see, what they share, and what remains sacred within. By choosing awareness over anxiety, we rediscover a truth as old as humanity: privacy is not about secrecy; it is about the dignity of being human.

CHAPTER 6

SOCIAL MEDIA AND DIGITAL FOOTPRINTS

When the internet first emerged, few lawmakers imagined how deeply it would reshape society. Early online spaces were open, unregulated, and built on the idea of freedom, the freedom to publish, to communicate, and to explore. As digital technology expanded into banking, healthcare, commerce, and social life, that freedom brought new challenges: privacy breaches, cybercrime, and the misuse of personal data.

Revisiting my own older posts while writing this section was a reminder that the internet rarely forgets, even when we do.

Governments realized that the same tools that connected people also exposed them. The first generation of digital laws, written in the 1990s and early 2000s, focused on issues such as fraud, copyright protection, and child safety. But the rise of social media, smartphones, and artificial intelligence transformed the problem from individual incidents to systemic risk.

Today, digital law touches every part of life. It defines how data is stored, how companies operate, how citizens are tracked, and how power is exercised. The conversation has shifted from "how do we regulate technology?" to "how do we preserve democracy, fairness, and privacy in a world run by data?"

Artificial intelligence makes this even more urgent. Governments can use AI for border control, surveillance, and public health. Corporations use it for advertising and prediction. Citizens rarely know where their data goes or who controls it. This lack of transparency is not just a legal issue; it is a moral one. Digital rights are quickly becoming human rights.

The Patchwork Problem

Despite the global nature of technology, digital laws remain largely national. Each country has its own approach to data protection, cybersecurity, and online regulation. The result is a fragmented legal landscape, a patchwork of rules that overlap, conflict, or leave gaps large enough to be exploited.

The European Union has adopted one of the most assertive stances through the General Data Protection Regulation (GDPR), a comprehensive law that affords individuals control over their personal data. Under the GDPR, individuals have the right to access, rectify, or delete their personal data held by companies. Businesses that fail to comply face enormous fines.

The United States relies heavily on self-regulation. Instead of a single federal privacy law, the U.S. employs a set of sector-specific rules—for healthcare (HIPAA), finance (GLBA), and children's data (COPPA). States such as California have enacted stronger protections, but these vary across jurisdictions. For multinational companies, compliance becomes increasingly complex, and for consumers, protection often depends on their geographical location.

In other regions, the situation is even less consistent. Some countries have minimal oversight, while others use technology as a means of control rather than protection. The absence of universal standards allows corporations to select the most favorable jurisdictions, thereby hosting data in countries with the weakest regulatory frameworks.

The internet is global, but the law is local. Until nations cooperate, privacy and accountability will remain uneven and uncertain.

Power and Responsibility

Governments occupy a challenging position in the digital age. They are responsible for protecting citizens, but they also have the power, and sometimes the temptation, to monitor them. Surveillance technology, justified by national security or crime prevention, can easily cross the line into a violation of personal freedom.

After the terrorist attacks of the early 2000s, many nations expanded their surveillance powers. Programs designed to detect threats began collecting massive quantities of data on ordinary citizens. Emails, phone calls, and browsing histories became part of vast digital archives. These measures, though often well-intentioned, blurred the boundary between protection and intrusion.

Artificial intelligence adds an additional layer. Machine learning systems can analyze patterns across billions of data points, predicting behaviors and identifying "risks." This predictive surveillance can prevent crime, but it can also unfairly target individuals based on biased data. Minority communities are especially vulnerable to such automated profiling.

The challenge is to strike a balance between safety and liberty. Laws must allow governments to act against legitimate threats while preventing abuse of power. Transparency and oversight are essential. Citizens should know how surveillance programs operate, what data they collect, and how long they keep that data.

The core question is timeless: Who watches the watchers? In a digital world, the answer must include public accountability, ethical design of technology, and informed participation.

Corporate Influence

Technology companies hold power once reserved for governments. Their platforms shape elections, economies, and culture. Their databases contain information about billions of people. The policies of corporations like Google, Meta, Amazon, and TikTok have a more direct impact on daily life than national laws.

Corporations claim to offer free services, but the actual cost is the collection of personal data. Every search query, location ping, and online purchase becomes part of a vast system of behavioral prediction. These companies profit from surveillance, not necessarily by selling data itself, but by selling access to attention.

Their influence extends to lawmaking. Through lobbying, donations, and partnerships, large tech firms help draft the very regulations meant to limit them. Sometimes, they propose voluntary "ethics principles" that appear responsible but lack enforcement mechanisms.

Governments often struggle to keep pace. The pace of technological innovation outstrips that of legislative reform. Lawmakers face pressure not to stifle innovation, which can lead to compromises that favor industry interests. The result is a growing imbalance between corporate power and public oversight.

Artificial intelligence magnifies this imbalance. The algorithms that power digital economies are proprietary secrets, protected by intellectual property laws. Even regulators cannot always examine how they work. Without transparency, accountability remains limited.

The solution lies not in hostility toward technology companies but in redefining their responsibilities. Innovation should not excuse exploitation. Corporations that benefit from public data must also protect it, not as a gesture of goodwill, but as a matter of law.

The Loophole Economy

Where laws exist, loopholes follow. In the digital world, those loopholes are both legal and technical. Companies, advertisers, and even governments exploit gaps between jurisdictions, definitions, and enforcement.

One typical example involves *data localization*. Many nations require that personal data collected within their borders remain on local servers. However, global companies route data through complex supply chains, using subsidiaries and cloud providers in various countries. The data never stays in a single place, making it nearly impossible to determine which laws apply.

Another loophole involves consent. Websites often ask users to accept privacy policies, but these documents are frequently lengthy and unclear. Consent got through confusion is not genuine, yet it remains legally valid in many systems. As a result, companies can claim compliance while users remain uninformed.

Artificial intelligence creates alternative forms of loopholes. Machine learning models trained on personal data can generate insights without technically "storing" identifiable information. This allows companies to argue that they are not violating privacy laws while still benefiting from patterns extracted from private data.

The loophole economy thrives on ambiguity. As long as technology grows faster than regulation, exploitation will continue. Closing these gaps requires not only new legislation but new definitions of privacy, ownership, and consent.

Global Perspectives

Different regions of the world offer contrasting lessons on balancing innovation with protection.

The European Union remains the global leader in data protection and privacy law. Data protection standards are set by the GDPR, which requires companies worldwide to adopt stricter measures. The EU is also advancing the **AI Act**, the first comprehensive legal framework governing artificial intelligence. It classifies AI applications by risk and imposes strict rules on systems used in law enforcement, healthcare, and employment.

The United States continues to rely on market-driven solutions. While agencies such as the Federal Trade Commission investigate deceptive data practices, there is no universal federal privacy law. State-level initiatives, such as California's Consumer Privacy Act (CCPA), partially fill the gap but leave inconsistencies. Tech companies often shape the conversation by defining their own privacy standards.

China offers a different model, one built on state control. The government integrates AI into governance, using facial recognition, social credit scoring, and extensive data monitoring. Although China has enacted data privacy laws, these prioritize national security and social order over individual autonomy.

Developing nations face a dual challenge: they seek economic growth through technology, but lack the resources to enforce robust privacy protections. Many rely on imported systems built by foreign companies, raising concerns about digital dependency and data colonialism—the extraction of information from poorer countries for the benefit of wealthier ones.

The global picture reveals a digital divide not only in access but in rights. While some citizens live under robust protections, others navigate online spaces with little or no legal safety net. True digital justice requires cooperation beyond borders.

Building a Fair Digital Future

The future of digital law depends on three forces: cooperation, education, and ethics.

Cooperation must replace fragmentation. Nations need shared principles for privacy, security, and accountability. Just as international treaties govern trade and the environment, the digital world requires its own charter of rights—one that protects individuals regardless of their geographic location. Organizations such as the United Nations and the World Economic Forum have started discussions on digital ethics, but progress has been slow. Binding agreements, not voluntary pledges, will define success.

Education is equally essential. Citizens cannot defend rights they do not understand. Digital literacy should include knowledge of how data flows, how AI systems make decisions, and how laws protect or fail to protect individuals. Informed people can hold governments and corporations accountable.

Ethics must guide innovation. Law can punish violations, but ethics prevents them. Designers, engineers, and policymakers should view privacy as a default, not a burden. Building ethical AI involves embedding fairness and transparency into the code from the outset.

There are promising signs. Countries are appointing digital rights commissioners. Activists push for "data dignity," the idea that individuals own the value their data creates. Young innovators design privacy-friendly alternatives to mainstream platforms. The conversation is shifting from fear to empowerment.

Ultimately, digital law is not about restricting technology but about aligning it with human values. Just as industrial societies created labor laws to protect workers, digital societies must create data laws to protect citizens.

Artificial intelligence and the global internet will continue to grow, but so can our systems of justice. A fair digital future depends not only on new regulations but on a shared understanding: that technology serves humanity best when it respects freedom, fairness, and dignity.

The challenge ahead is not whether we can build more intelligent machines, but whether we can develop wiser laws. The future of privacy and indeed democracy itself will depend on how well we rise to this task.

CHAPTER 7

THE CONCEPT OF DIGITAL HYGIENE

This chapter stayed with me long after I finished drafting it, because the stakes here feel deeply personal, not theoretical. When most people think of hygiene, they typically envision brushing their teeth, washing their hands, or keeping their home organized. These habits promote physical health and foster a sense of peace of mind. But in the modern era, another form of hygiene has become equally essential — *digital hygiene*.

Digital hygiene refers to the routine practices that keep our online lives safe, organized, and efficient. It includes updating passwords, clearing cluttered inboxes, managing privacy settings, and maintaining awareness of what personal information is being shared. Just as physical hygiene prevents illness, digital hygiene prevents data breaches, identity theft, and emotional burnout.

The term might sound technical, but it is deeply human. Digital hygiene is about control, regaining ownership of our digital environments so that they serve us, not the other way around. It encourages reflection: how much time do we spend online, and how much of our data have we unknowingly surrendered?

As neglecting physical hygiene can lead to illness, neglecting digital hygiene can allow vulnerabilities to accumulate. Forgotten accounts become entry points for hackers. Unsecured Wi-Fi networks expose private communications. Outdated software invites malware. Even the endless buildup of photos, emails, and files can slow down devices and mental focus alike.

The foundation of digital hygiene is mindfulness, an awareness of what we do, why we do it, and how it affects our security and well-being.

The Hidden Mess Behind the Screen

Every device we use contains traces of our lives. Emails, downloads, old messages, photos, and search histories all leave behind digital fingerprints. Over time, these fragments accumulate into a hidden clutter that mirrors the mess of an unorganized home.

Unlike physical clutter, however, digital mess often goes unnoticed. Files take up invisible space on servers rather than on tabletops. Apps run quietly in the background, collecting data and draining energy. Social media profiles, even those rarely used, continue to store personal information indefinitely.

Each forgotten password, old account, and unchecked setting represents a potential risk. Hackers exploit outdated data and inactive profiles because they are less likely to be monitored. A decade-old account on a forgotten website may still contain personal information that links back to current identities.

The digital mess extends beyond devices. Our data lives in the cloud, on servers scattered across the world. Many people assume that deleting a file permanently removes it, but in reality, backups, shared drives, and synced platforms can preserve that information for years.

Digital hygiene begins by confronting this invisible mess. Cleaning it up means asking difficult questions: What information truly needs to exist? What can be deleted, archived, or secured? It also means facing the emotional side of clutter — the unease of letting go of digital remnants tied to memory or habit.

Good digital hygiene restores order, clarity, and safety. It reduces not only risk but also mental noise.

Security Starts with Simplicity

At its core, digital hygiene is about prevention, not perfection. Just as washing hands regularly is more effective than elaborate post-infection cures, small, consistent habits protect against most digital threats.

Strong Passwords and Authentication

Weak passwords remain one of the most common causes of data breaches. Many people reuse the same password across multiple sites, making it easy for hackers to exploit a single leak. A strong password should be long, unique, and complex, ideally a phrase rather than a single word. Password managers simplify this process by generating and storing secure credentials.

Two-factor authentication adds an extra layer of safety by requiring a second form of verification, usually a code sent to a phone or an app. Even if someone steals a password, the system still protects the account.

Regular Updates and Patches

Outdated software is a favorite target for cybercriminals. Updates may seem inconvenient, but they often fix vulnerabilities before someone can exploit them. Maintaining current versions of apps, operating systems, and browsers is one of the simplest and most effective ways to defend against security threats.

Backups and Recovery

Accidents happen. Devices fail, files get corrupted, or ransomware locks access. Regular backups, securely stored and encrypted, enable the recovery of valuable data. Cloud storage simplifies this, but you should verify where and how the data is stored.

Privacy Settings and Permissions

Every app and website requests access to information. Many of these permissions are unnecessary. Reviewing them regularly helps prevent overexposure. Turning off location tracking, limiting ad personalization, and revoking unused app access are practical steps toward digital hygiene.

Simplicity does not mean weakness. It means consistency: building small, sustainable habits that protect privacy without overwhelming effort.

The Psychological Side of Digital Clutter

Clutter has psychological consequences, whether physical or digital. Inboxes filled with thousands of unread messages, endless notifications, and disorganized files create mental static. Each ping demands attention, fragmenting focus and raising anxiety.

Research shows that multitasking with constant digital interruptions can reduce productivity and satisfaction. The brain, forced to switch rapidly between tasks, tires quickly and processes information less deeply. This leads to what researchers call *cognitive fatigue*.

Digital clutter also carries emotional weight. Old photos, messages, or unfinished projects can trigger nostalgia, guilt, or distraction. In an era where memory is outsourced to technology, deleting files can feel like erasing part of one's identity. Yet just as minimalism in physical spaces brings calm, reducing digital excess can bring clarity.

The psychological effects extend to self-image. Social media platforms encourage constant engagement, rewarding visibility and speed. As our digital environment becomes increasingly cluttered, it becomes harder to distinguish between personal goals and algorithmic expectations. Digital hygiene reintroduces intention, the choice to engage thoughtfully rather than reactively.

Cleaning up devices and feeds is not just a technical act; it is an act of mental health. It signals that we value our time and attention.

Practical Clean-Up Strategies

Improving digital hygiene does not require technical expertise; it only requires awareness and routine. The following steps help restore order and reduce risk:

1. Audit Your Accounts

List every online account you have ever created. Delete or deactivate those you no longer use. Many old accounts continue to collect and share data without your knowledge.

2. Review Privacy Settings

Check the privacy controls on social media, browsers, and mobile devices. Restrict data sharing to what is necessary. Opt out of targeted advertising when possible.

3. Clean Your Inbox and Files

Unsubscribe from newsletters you never read. Delete spam and unnecessary attachments. Organize essential documents into folders, and back them up.

4. Manage Your Devices

Remove unused apps. Check for updates regularly. Secure devices with strong passwords or biometrics. Avoid connecting to unsecured public Wi-Fi networks.

5. Monitor Your Digital Footprint

Search your name online to see what information is publicly visible. Request removal from data broker sites if necessary. Tools like "Have I Been Pwned" can alert you to breaches involving your email addresses.

6. Schedule Regular Maintenance

Just as homes need cleaning and cars need service, digital environments require attention. Set monthly reminders to review accounts, passwords, and privacy settings.

The goal is not perfection but progress. Each small improvement strengthens digital resilience and fosters confidence.

These steps reflect what I've implemented myself, gradually, imperfectly, and with adjustments along the way,

Digital Balance and Boundaries

Digital hygiene extends beyond security; it includes emotional balance and boundaries. In a hyperconnected world, constant notifications blur the line between work, rest, and reflection. The brain rarely gets a chance to reset.

Setting boundaries begins with awareness. Notice how often you check your phone or refresh a feed. Ask whether these habits serve a purpose or simply fill the silence. Turning off non-essential notifications and scheduling device-free hours restores focus and calm.

Workplaces are also beginning to recognize the value of digital balance. Some countries have implemented "right to disconnect" laws, allowing employees to ignore emails and messages outside working hours. This approach acknowledges that mental health and digital boundaries are intertwined.

Creating personal boundaries might mean setting screen time limits, silencing alerts at night, or designating "tech-free zones" in the home. It also means nurturing offline connections, such as conversations, hobbies, and time spent in nature.

Digital balance is not about rejection; it is about rhythm. The healthiest relationships with technology involve both use and pause, as well as connection and reflection.

Healthy Habits for the Digital Age

Maintaining digital hygiene is not a one-time project but an ongoing practice. Like exercise or nutrition, it becomes easier when integrated into daily life.

Start Small and Stay Consistent

Delete one unused app each week. Clean a single folder each day. Over time, these habits accumulate into lasting order.

Protect Personal Data as You Would Cash or Keys

Think of your information as valuable property. Before sharing, ask: "Would I give this to a stranger?" Awareness prevents impulsive clicks and careless sign-ups.

Use Technology Intentionally

Before downloading a new app or joining a new platform, ask what purpose it serves. Does it make life easier, or simply add noise? Digital minimalism is a form of self-respect.

Educate and Empower Others

Share digital hygiene practices with friends, family, and children. Encourage conversations about privacy, consent, and online safety. A culture of awareness multiplies protection.

Revisit and Reflect

Once or twice a year, take a "digital retreat." Review habits, declutter accounts, and reset priorities. Reflection keeps behavior aligned with values.

Healthy digital habits build confidence and peace of mind. They transform technology from a source of anxiety into a tool for empowerment.

The digital world is vast, powerful, and constantly changing. Yet, within it, individuals still hold the power of choice, the ability to control attention, manage information, and protect their identities.

Digital hygiene is not about fear or perfectionism. It is about balance, responsibility, and care. When practiced regularly, it creates a cleaner, safer, and more focused digital life, one that supports rather than consumes us.

In the end, just as clean hands prevent illness, clean habits preserve clarity. The tools we hold in our hands every day — phones, tablets, laptops — are extensions of our minds. Keeping them organized and secure is not just smart; it is essential for living wisely in the digital age.

CHAPTER 8

SURVEILLANCE, CONVENIENCE AND CONSENT

In just two decades, a handful of technology companies have grown from garage startups to global empires. Google, Apple, Amazon, Meta, Microsoft, and a few others dominate the digital landscape, shaping how we communicate, shop, learn, and even think. Their platforms connect billions of people and process unimaginable amounts of data every second.

These companies have changed the world, often for the better. They have democratized access to information, accelerated medical research, connected families across continents, and created vast economic opportunity. Yet their influence also raises profound questions about power, privacy, and human autonomy.

Big Tech is not simply a collection of companies. It is an ecosystem, a network of interlocking services, devices, and algorithms that work together to gather, analyze, and monetize human behavior. The average person uses products from at least three of these companies daily, often without realizing the extent to which they interact. Smartphones send location data that fuels ad targeting. Search engines record queries that refine machine learning. Social media tracks engagement, which determines what content appears next.

This interconnected system forms what scholars call the *attention economy*. Time and focus have become commodities. They measure and monetize every click, scroll, and pause. The platforms that appear free are, in reality, funded by user behavior. The more time people spend on them, the more valuable their data becomes.

The rise of Big Tech has transformed not only markets but also the very structure of modern life. To tame this ecosystem, one must first understand its mechanisms, how convenience conceals complexity, and how algorithms quietly shape the world behind every screen.

What became clear to me is how often consent is assumed rather than clearly given.

The Power of Platforms

Platforms are the digital environments where modern life unfolds. They are the new public squares, marketplaces, and classrooms of the 21st century. Yet unlike traditional public spaces, they are owned and controlled by private corporations. Their rules, not democratic laws, determine what is visible, permissible, and profitable.

Algorithms are at the heart of this power. Designed to optimize engagement, they decide what posts, products, and headlines appear first. They tailor feeds to individual preferences, creating a personalized digital universe for each user. On the surface, this seems efficient. Who wouldn't want relevant recommendations? But personalization comes with hidden consequences.

Algorithms are not neutral. They reflect the values of those who design them and the biases inherent in the data they collect. Their primary goal is not truth or fairness but attention. If outrage or fear keeps users clicking longer, those emotions become the raw material of profit.

This influence extends beyond entertainment. Algorithms can sway elections, shape public discourse, and affect self-esteem. The content one sees online depends not only on personal choice but on invisible calculations that prioritize engagement over understanding.

The power of platforms lies in this invisibility. Unlike governments or media outlets, they operate with limited transparency. Their terms of service change frequently, often without notice. Their moderation policies determine what speech is allowed to survive online. For many, the digital world has become the primary lens through which reality is viewed, a lens crafted by corporate interests and optimized for revenue.

Recognizing this power is the first step toward reclaiming it. Users who understand how platforms shape their perceptions can begin to navigate them with awareness rather than assumption.

Data as Currency

In the digital age, data has replaced oil as the world's most valuable resource. Every search query, purchase, and location ping adds to the growing data economy. Unlike traditional commodities, data is renewable. Every action generates more.

Big Tech companies collect, process, and trade this information on an unprecedented scale. They use it to build behavioral profiles that predict future choices, from shopping preferences to political leanings. These predictions fuel targeted advertising, which funds the "free" services people rely on.

Most users are vaguely aware that their data is being collected, but underestimate the extent of its collection. Smartphones record daily movement patterns, voice assistants capture spoken commands, and wearable devices monitor heart rates and sleep cycles. Even offline behavior can be inferred through online patterns.

This data does not simply describe behavior. It influences it. By predicting preferences, companies can also steer them. When an ad appears at the right moment, it can subtly influence a decision before the user even realizes it.

The danger is not only the loss of privacy but also the concentration of power. When a few companies control most global data, they gain unprecedented insight into society. They can shape markets, influence culture, and anticipate trends before governments or citizens can respond.

Data is the currency of the modern world, but users rarely see its value. The more aware individuals become of its worth, the more likely they are to demand fair exchange — services that respect, rather than exploit, their information.

The Hidden Costs of Convenience

Every tap and swipe in the digital world promises ease. Online shopping replaces travel, streaming replaces television schedules, and social media replaces community gatherings. Yet convenience often conceals a deeper cost.

When everything is effortless, awareness diminishes. People often trade privacy for speed without realizing it. Auto-login features, saved passwords, and personalized recommendations simplify life but also surrender control. Each shortcut hands another fragment of personal data to unseen systems.

The psychological cost is equally real. Constant convenience encourages impulsivity. With one click, a product arrives. With one tap, a video plays. The brain receives instant reward, reinforcing habits that prioritize gratification over reflection. Over time, attention spans shorten, and patience often erodes.

Also, the environmental cost is significant. Big Tech's infrastructure, along with constant device upgrades, consumes an enormous amount of energy. The "cloud" is not weightless; it is a network of physical servers that require electricity, cooling, and rare materials.

Convenience is not inherently negative. It becomes harmful only when it removes choice and awareness. The goal is not to reject comfort but to understand its price. True digital empowerment lies in deliberate use of technology, not in automatic use.

Manipulation by Design

Big Tech's most powerful aspect is not its algorithms or its data, but its design, which shapes behavior through interface construction. Designers meticulously craft every color, sound, and notification to capture attention. This practice, known as *behavioral design*, integrates psychology and technology to maintain user engagement.

For example, social media platforms use intermittent rewards. This is the same principle that drives slot machines. Each refresh of a feed presents an opportunity for something new and exciting. The unpredictability keeps users scrolling. Likes and notifications trigger dopamine releases, creating pleasure loops that mimic addiction.

Streaming services auto-play the next episode to reduce decision-making friction. E-commerce platforms use scarcity tactics, such as "only two left in stock," to create a sense of urgency. Even email apps nudge users with prompts to reply or schedule meetings. These subtle designs encourage constant interaction.

The ethical problem is that users rarely consent to such manipulation. The design of digital platforms shapes not only behavior but also emotion and identity. It blurs the line between persuasion and control.

Awareness is the antidote. By recognizing how design influences attention, individuals can develop strategies to resist compulsive patterns. Turning off auto-play, muting notifications, or setting screen time limits are simple ways to reclaim your agency. Behavioral design works both ways, and it can also build healthier, more intentional habits.

Breaking the Cycle

The Big Tech ecosystem thrives on dependence. More data gets collected, and the algorithms become more refined as people spend more time online. It is a self-reinforcing cycle that rewards attention and punishes absence.

Breaking the cycle does not mean abandoning technology but redefining its role. It begins with conscious disengagement: taking deliberate pauses from screens, disabling nonessential notifications, and questioning default settings.

Practical steps help restore balance:

- **Diversify your platforms.** Use independent search engines, privacy-focused browsers, and smaller social networks when possible.
- **Limit algorithmic control.** Switch from "recommended" content to chronological feeds. Choose what to follow rather than letting algorithms dictate your choices.
- **Reclaim ownership of data.** Delete unused accounts, clear cookies, and use privacy tools that minimize tracking.
- **Create friction intentionally.** Log out after sessions or avoid saving payment information. Minor inconveniences slow down impulse decisions and increase awareness.
- **Support ethical technology.** Choose companies that value transparency, sustainability, and user rights.

The goal is not disconnection but discernment. By understanding how Big Tech captures attention, users can choose when and how to engage. This transformation from passive consumer to active participant is the essence of digital empowerment.

Taking Back Control

Taming the Big Tech ecosystem begins with awareness, but it

ultimately requires collective and individual action.

At the personal level, individuals can cultivate digital mindfulness. This means being intentional with time, curating what information enters one's mind, and setting boundaries that protect attention. It means reading before sharing, pausing before reacting, and choosing depth over distraction.

On a societal level, accountability must accompany innovation. Governments can enforce transparency in algorithmic systems, ensuring that machine-made decisions are explainable and fair. Data privacy laws must keep pace with technological developments, ensuring that users have the right to access and understand how their data is used.

Education plays an equally vital role. Digital literacy should encompass not only the use of technology but also the critical questioning of it. Schools and communities can teach critical awareness of algorithms, advertising, and online influence, skills as essential as reading and writing.

Finally, ethics must guide progress. Engineers, designers, and leaders bear responsibility for the systems they create. Technology should enhance human dignity, not diminish it. Profit should never outweigh privacy, and innovation should never come at the expense of trust.

The digital future does not belong solely to corporations. It belongs to the billions of people who use, shape, and depend on these systems daily. Taming Big Tech is not an act of rebellion; it is an act of reclamation, a reassertion of human agency in an age of automation.

Technology is a tool, not a destiny. When used wisely, it can amplify creativity, connection, and understanding. But it must serve humanity, not control it. The task ahead is to ensure that the digital ecosystem grows not only larger and smarter but also kinder and more just.

Each individual who learns to navigate this world with awareness contributes to that balance. Every thoughtful click, pause, or refusal to share personal data is a minor act of resistance against manipulation, and a step toward a healthier digital civilization.

The world of Big Tech may be vast, but human intelligence, empathy, and integrity remain more powerful. By using these qualities as our compass, we can navigate the digital landscape with confidence and reclaim control of our connected lives.

CHAPTER 9

PROTECTING YOURSELF WITHOUT BECOMING PARANOID

The workplace and the classroom have long been places of structure and observation, but digital technology has transformed them into environments of near-constant visibility. What began as tools for productivity and learning has evolved into systems for tracking, measuring, and predicting human behavior.

Cameras monitor offices and hallways. Software tracks keystrokes, attendance, and online activity. Learning platforms record every click, assignment, and login time. In both work and school, privacy has become a scarce commodity.

This surveillance is often justified in the name of efficiency and safety. Employers claim that monitoring ensures accountability. Schools argue that data collection improves educational outcomes. However, these tools can erode trust, autonomy, and creativity, the very qualities that enable people to thrive.

For many, this shift feels subtle. It arrives through everyday convenience: digital badges, smartboards, learning analytics, and performance dashboards. Yet behind these technologies lies a growing expectation that every moment should be measurable.

As the lines between personal and professional life blur, individuals confront a new challenge: safeguarding their dignity in spaces where others might record, analyze, and evaluate their every action.

I don't believe constant fear leads to better choices, but informed awareness has changed the way I use technology every day.

Productivity or Control?

The modern obsession with data has turned workplaces and schools into laboratories of constant measurement. People can quantify every task and log every interaction. The intent is often practical, to increase productivity or ensure fairness, but the outcome is more complex.

In offices, productivity tools record screen time, email response rates, and meeting attendance. Some companies use webcams or motion sensors to monitor remote workers. Algorithms evaluate output, ranking employees on performance dashboards.

In schools, especially after the rise of remote learning, digital tracking has become routine. Attendance software records login times, plagiarism tools scan student writing, and educational platforms measure "engagement" based on how long a student's eyes remain on the screen.

The line between helpful feedback and invasive scrutiny grows thin. Constant measurement creates a culture of surveillance rather than support. Instead of motivating improvement, it breeds anxiety and self-censorship. People begin to perform for the system rather than engage authentically.

When every moment is observed, people reduce their risk-taking. Creativity suffers. Students become afraid to experiment or take risks, fearing failure. Workers hesitate to propose unconventional ideas. The result is a paradox: tools designed to enhance productivity often reduce the very innovation that drives progress.

True efficiency depends on trust, not control. When people feel safe and respected, they contribute more openly and think more boldly. When they feel monitored, they conform.

Digital Footprints at Work

Few employees realize how much data their workplaces collect about them. Beyond basic records such as payroll and attendance, companies collect data on digital activity, communication patterns, and even physical movement.

Companies often archive emails, chat messages, and video calls indefinitely. Security systems log entry and exit times. Network tools monitor which websites users visit. In some industries, wearable devices monitor biometric data, such as heart rate or stress levels, and market themselves as wellness initiatives.

Employers justify this surveillance as necessary for safety, efficiency, or compliance. Yet the sheer volume of data collected raises serious questions about consent and control. Who owns this information? How long is it stored? Can it be used to make decisions about promotions or dismissals?

The problem is not always malicious intent. Many organizations adopt these technologies without considering their ethical or psychological implications. Data-driven management promises objectivity but can easily become impersonal. Algorithms cannot capture empathy, nuance, or personal circumstance.

Digital footprints extend beyond the workplace. Employers increasingly screen social media profiles, online portfolios, and public posts during the hiring process. A casual comment made years earlier can resurface as grounds for judgment. The professional and the personal have merged into a single, searchable identity.

To protect privacy at work, awareness is essential. Employees need to understand what data is collected, question unclear policies, and advocate for transparency. Privacy is not a privilege; it is a condition for genuine engagement and trust.

Privacy in the Classroom

The classroom, once a sanctuary of curiosity and growth, has also become a site of surveillance. Technology has redefined how students learn, how they are assessed, and how they are taught.

Digital learning platforms record every assignment submission, the time spent reading, and quiz scores. Cameras in remote learning sessions capture faces and environments. Proctoring software monitors eye movement and sound to detect "suspicious behavior." Even classroom attendance systems use facial recognition to log participation.

Schools and universities justify these tools as a means to ensure fairness, safety, and accountability. Yet the result can feel intrusive. Students say digital proctors make them feel anxious, as if they are presumed guilty until they prove their honesty.

The rise of educational analytics, the practice of analyzing vast datasets of student data, also raises ethical concerns. Advocates argue it helps personalize learning and identify students who need support. Critics warn it risks labeling students permanently based on early performance or behavior.

Children and young adults are vulnerable because they often cannot give informed consent. Institutions or corporations may retain data collected during their formative years for decades.

Authentic learning thrives in environments of trust and curiosity, not fear. When students feel constantly evaluated, they focus on avoiding mistakes rather than exploring new ideas. Digital tools should support education, not replace the human connection that gives it meaning.

Emotional and Psychological Costs

Being watched changes how people think, feel, and behave. The constant awareness of observation — even when indirect — alters the brain's sense of safety. Psychologists call this the *observer effect*: the mere belief that one is being monitored can cause anxiety, tension, and self-censorship.

In workplaces, employees under surveillance often report higher stress levels and lower morale. Employees report a sense of mistrust, feeling that employers value output over judgment or creativity. The pressure to appear productive can lead to burnout, as workers overextend themselves to meet metrics rather than meaningful goals.

In schools, surveillance creates similar psychological strain. Students under constant digital observation lose the freedom to make mistakes or express unpopular opinions. Instead of curiosity, they learn caution. The classroom becomes a stage rather than a workshop.

Long-term exposure to such conditions can lead to what researchers call *performance fatigue*, the exhaustion that comes from always being "on." Over time, this erodes confidence and reduces intrinsic motivation. People begin to define their self-worth through external metrics, such as grades, scores, or productivity charts, rather than genuine learning or growth.

Privacy is not the enemy of accountability; it is its foundation. When people have space to think without judgment, they can reflect, innovate, and improve. Without that space, authenticity gives way to compliance.

Your Rights and Responsibilities

Understanding rights in digital environments is the first step toward reclaiming privacy.

At Work

Many countries permit employers to monitor employees' devices, email, and networks legally. However, transparency is required in most jurisdictions. Employers should inform employees about what they observe and why. Some regions, such as the European Union, offer stronger protections under laws like the General Data Protection Regulation (GDPR), which grants individuals the right to access and correct their data.

Workers also have responsibilities. Using strong passwords, avoiding personal activities on work devices, and carefully reading company policies can prevent misunderstandings and reduce risk. Advocacy within organizations, through employee committees or unions, can also help establish fair data practices.

At School

Many countries protect students, especially minors, under education and privacy laws. In the United States, the Family Educational Rights and Privacy Act (FERPA) restricts access to student records. In other regions, similar laws require parental consent before third parties may access the data.

However, technology often advances more rapidly than regulation. Parents, teachers, and students must work together to ensure that educational software providers uphold ethical standards. Schools should make public the data they collect, how long they keep it, and whether they share it with vendors.

The most effective protection still comes from awareness. Knowing the rules allows individuals to assert their rights and hold institutions accountable.

Reclaiming Dignity in Shared Spaces

Digital monitoring in workplaces and schools will not disappear entirely. Safety, fairness, and accountability are legitimate goals. But there is a profound difference between supervision that supports and surveillance that controls. The challenge is to build environments that respect both structure and humanity.

Reclaiming dignity begins with open conversation. Employers can discuss with workers how they use monitoring data. Schools can involve students and parents in decisions regarding the adoption of technology. Transparency transforms suspicion into trust.

Technology itself can also be designed with ethics in mind. Privacy-enhancing tools, anonymized data collection, and limited retention periods can achieve accountability without intrusion. Innovation and protection need not conflict.

Individuals can also reclaim power through their daily choices. Logging out of shared accounts, questioning unnecessary data requests, and maintaining separate devices for personal and professional use are simple acts of boundary-setting. These actions signal self-respect and responsibility.

Cultural change takes time, but it begins with awareness. The more people understand how surveillance shapes behavior, the more they can advocate for systems that prioritize well-being over metrics.

Work and education are not just about performance; they are about purpose. We should nurture human potential instead of exhausting it. The future of privacy depends on re-centering these institutions around trust, compassion, and respect for individuality.

To protect privacy is to protect humanity — the right to think freely, learn honestly, and grow without fear. The digital tools surrounding modern life should amplify these values, not erode them.

By demanding transparency, practicing digital mindfulness, and fostering empathy in shared spaces, we can restore balance. Surveillance may watch what we do, but it cannot define who we are. When dignity guides technology, privacy becomes not a privilege, but a shared right, one that strengthens both community and character.

CHAPTER 10

PRACTICAL STEPS YOU CAN TAKE NOW

In earlier generations, identity evolved through experience, which was shaped by family, culture, community, and memory. Today, it is increasingly mediated through screens. The person we see in the mirror is only one version among many: our physical self, our digital self, and the curated image that others perceive through algorithms and platforms.

Every post, comment, and "like" becomes a digital brushstroke, painting the picture of who we appear to be. Yet this identity is not static. It changes depending on the audience and the platform: professional on LinkedIn, humorous on TikTok, idealized on Instagram, and private on encrypted apps. The modern self has become fragmented, split across virtual spaces, each demanding its own version of authenticity.

This fragmentation is not inherently artificial. Humans have always shown different sides in different settings—the professional at work, the parent at home, the friend in conversation. But the scale and permanence of digital visibility have altered the balance. Online, people can preserve, reshare, and judge every moment. The boundaries that once separated public and private life have blurred into a continuous performance.

This new reality invites both freedom and pressure. People can express parts of themselves that once had no outlet, connecting with others who share their values or struggles. Yet the same tools that enable connection also create distortion. Identity becomes an ongoing negotiation between expression and expectation, between who we are and who algorithms think we should be.

Avatars and Algorithms

An *avatar* was once a symbol of transformation, a representation of one's spirit in another realm. In the digital age, it has become literal: the profile photo, the username, the personal brand. These digital stand-ins carry our reputations, our work, and our relationships. They have become essential for participation in modern life.

But avatars are not neutral. The platforms that host them and the algorithms that amplify them shape them. Each interaction, a like, a share, a comment, feeds into a feedback loop that trains algorithms to define what kind of person we are. Over time, the algorithm learns our preferences and reinforces them, narrowing our view of the world to match what we already engage with.

This process creates a paradox of personalization. The more an algorithm "understands" us, the less we see beyond ourselves. The news, entertainment, and advertisements that appear on our screens are not reflections of objective truth but of our predicted interests. In this way, algorithms don't just observe identity; they construct it.

Social media amplifies this loop through design. Metrics such as followers, likes, and impressions become external measures of self-worth. Each notification serves as micro-validation, encouraging the repetition of behaviors that attract attention. Over time, many people find themselves presenting versions of their identity that please algorithms rather than reflecting their inner truth.

The result is a new kind of digital echo chamber — not only political or informational but personal. We see the world through the mirror of what our past clicks suggest we want. In this environment, self-discovery competes with self-promotion, and genuine individuality risks becoming just another algorithmic pattern.

The Illusion of Connection

Technology has made connection effortless but not necessarily meaningful. Networks surround us, yet loneliness has become one of the defining emotions of the digital age. The same tools that link billions of people often leave individuals feeling unseen.

Part of the reason lies in how digital platforms frame interaction. Communication online is immediate, measurable, and public. Reaction icons and character limits replace the human nuances of tone, eye contact, and silence. Relationships become streams of data points: likes, comments, messages, streaks. Quantity replaces quality.

For students and professionals alike, this creates emotional dissonance. People may have hundreds of contacts but few confidants. Validation becomes transactional, dependent on visibility rather than intimacy. The pressure to remain "connected" at all times breeds exhaustion and comparison, rather than contentment.

The illusion of connection is powerful because it offers partial satisfaction. It mimics belonging without requiring vulnerability. Yet authentic connection, whether at work, in school, or among friends, depends on presence and trust. It cannot thrive under constant performance pressure.

Genuine digital connection begins when we use technology intentionally rather than reactively. When we listen as much as we broadcast, when we pause before we post, and when we treat online spaces as extensions of empathy rather than stages for approval.

From Data to Definition

Data is no longer just a record of what we do; it has become part of who we are. Companies collect, categorize, and predict based on digital behavior, turning personality into a measurable asset. In this world, a person's identity is not only a reflection of their choices but also a product of the systems that interpret them.

Every online action adds to an invisible file, a behavioral biography that defines how others see us. Credit agencies use it to assess reliability, employers to evaluate potential candidates, and advertisers to estimate demand. The data we generate becomes a kind of external memory, one that others can read and analyze.

This externalization of identity creates both empowerment and vulnerability. On one hand, data helps personalize experiences, from playlists that anticipate mood to navigation apps that predict routes. On the other hand, it introduces asymmetry. The systems that know us so well rarely share what they know. Our identities, shaped by their interpretations, are fed back to us as recommendations, rankings, and reputations.

This process raises philosophical questions: who owns the modern self? If identity can be quantified, who decides its boundaries? When a person's data history determines their opportunities, how free are they to change?

To navigate this future, we must separate identity from its digital residue. Data describes behavior but does not define humanity. Identity is not a set of analytics; it is the ongoing story we tell about who we are, a story that no algorithm can fully capture.

The Crisis of Authenticity

Authenticity, once considered a personal virtue, has become a performance. The digital world rewards constant visibility — the curated image, the polished narrative, the perfect caption. In this environment, being authentic is often conflated with being relatable in a marketable way.

The constant exposure reshapes self-perception. Many begin to view their lives through the lens of potential content, evaluating experiences based on how they appear online rather than how they truly feel. This self-surveillance creates a double consciousness, living both as oneself and as the imagined audience watching.

The consequences are both emotional and ethical. Studies show that constant comparison on social media contributes to anxiety, depression, and lowered self-esteem. The drive to appear authentic can paradoxically produce conformity, as people emulate the styles, opinions, and aesthetics that attract the most engagement.

For young people growing up in this environment, the pressure to maintain multiple digital identities can be overwhelming. Mistakes that once faded with time now persist in searchable archives. The past, preserved online, can feel like a shadow that never disappears.

Authenticity cannot survive in an economy that rewards imitation. It requires freedom — the right to change, to be inconsistent, to grow without being algorithmically penalized for it. The future of digital identity depends on reclaiming that freedom.

Building a New Digital Self

If identity in the digital age is fragmented, performative, and data-driven, then rebuilding it requires intention and ethics. The challenge is not to reject technology but to design and use it in ways that preserve human integrity.

Digital literacy must expand beyond technical skills to include emotional and ethical awareness. People need to understand not just how to use technology, but how technology uses them. This includes recognizing when design manipulates attention, when personalization limits perspective, and when data collection exceeds consent.

Transparency and consent should become the foundation of digital systems. Users must be able to see, understand, and control what data defines them. Regulations such as the European Union's GDPR and California's Consumer Privacy Act are steps in this direction, but cultural change is equally important. Awareness must become a social norm.

Design ethics can play a transformative role. Platforms can prioritize well-being over engagement, allowing people to curate identity without pressure. Features that promote reflection, such as time limits, context reminders, and content warnings, can help restore autonomy. Instead of optimizing for clicks, we can maximize technology for care.

Education is crucial. Schools and workplaces should teach identity resilience. This is the ability to navigate online environments without losing a sense of self. This includes encouraging students to distinguish between digital profiles and personal worth, and guiding employees to strike a balance between transparency and privacy.

Ultimately, building a new digital self requires humility. We must accept that the online version of us is only a fragment, a reflection. It's one voice in a larger conversation about what it means to be human.

Reclaiming the Human Story

As artificial intelligence, virtual reality, and biometric data reshape the boundaries of identity, one truth remains: technology cannot replace the essence of humanity. It can mirror it, magnify it, even manipulate it, but it cannot create meaning without human intention.

The future of identity depends on remembering that we are not merely data points or digital avatars. We are storytellers. The power to define who we are belongs not to algorithms but to the choices we make about how we use them.

Reclaiming the human story means living with awareness and compassion in digital spaces. It means questioning the systems that profit from our attention, resisting the urge to perform for validation, and using technology as a mirror for growth rather than as a stage for approval.

We can begin this reclamation in simple ways: by pausing before posting, by seeking connection over reaction, by protecting moments of offline silence where no one is watching. These acts restore the private self — the self that thinks freely, feels deeply, and exists beyond measurement.

In the coming years, identity will become even more fluid. Digital twins, virtual personas, and AI-generated representations will blur the line between self and simulation. The challenge will not be to stop these developments but to ensure they remain anchored to human values — empathy, truth, and dignity.

To reclaim the self in an algorithmic world is to remember that technology is a tool, not a mirror of worth. We shape it through the stories we tell and the boundaries we uphold. The more consciously we engage, the more humane our digital future becomes.

The ultimate question is not what algorithms will know about us, but what we will know about ourselves.

If we can look into the digital mirror and still recognize our reflection, which is flawed, complex, and developing, then identity will endure not as data, but as the timeless expression of what it means to be human.

I finished this chapter more aware of how much the future depends not on technology alone, but on the choices we're willing to make now.

CHAPTER 11

PROTECTING THOSE MOST AT RISK: WOMEN, CHILDREN, AND TEENS

The internet's purpose was to unite people, promote equality, and amplify the voices of those who were unheard. Yet, as with every technology, access does not guarantee equality. The digital world reflects and often amplifies existing divisions in society.

Women, children, older adults, and marginalized groups frequently face the highest risks and the fewest protections online.

The *digital divide* refers not only to unequal access to devices and networks but also to differences in digital literacy, awareness, and safety. In many parts of the world, women still have limited access to the internet, and they are discouraged from using technology. When they engage online, they often encounter harassment, exploitation, or surveillance that discourages participation.

Children, meanwhile, are born into a connected world they did not choose. Their digital footprints begin before they can speak, as parents share photos, schools collect data, and companies market directly to them. By adolescence, people can already track, analyze, and use many complex online identities for commercial or malicious purposes.

The digital divide is not only a matter of access; it is a matter of power. Those who understand technology control its use. Those who lack education or protection face the consequences of its misuse. To build a safer digital future, society must recognize that vulnerability in the virtual world often mirrors vulnerability in the physical one.

Technology can empower, but it can also exploit. The challenge is to ensure that the tools meant to connect humanity do not become instruments of control, manipulation, or harm, especially for those least equipped to defend themselves.

Women Online
Safety, Harassment, and Visibility

Women are among the most active users of social media and digital platforms, yet they are also among the most targeted. Online harassment, cyberstalking, and image-based abuse have become everyday experiences for many women around the world. What begins as digital aggression can spill over into physical danger, emotional trauma, and professional damage.

Online harassment based on gender often follows predictable patterns. Women who express opinions in public spaces, particularly about politics or culture, face disproportionate backlash. Trolls use anonymity to spread hate, send threats, and manipulate social media algorithms to silence or intimidate. Journalists, activists, and professionals report higher rates of digital abuse, including doxing, which is the public release of personal information.

Technology can magnify old patterns of discrimination. Deepfake videos, for example, use artificial intelligence to create fake sexual images, disproportionately targeting women. Revenge pornography, though illegal in many regions, continues to circulate because of weak enforcement and rapid content sharing.

The psychological toll is profound. Many women withdraw from public discourse or limit their visibility to avoid abuse. This self-censorship undermines diversity of thought and erodes equality in digital spaces.

Solutions require both awareness and accountability. Platforms must implement stronger reporting mechanisms, governments must enforce laws against online harassment, and users must learn to recognize and challenge toxic digital behavior. Education on digital ethics and empathy can foster cultural change, teaching that behind every username is a human being deserving of respect.

Women's visibility online should not be a liability. It should be a right, one supported by communities, institutions, and technologies designed to protect, not endanger.

Children in the Networked World

For children, the digital world is not an option; it is a reality. From streaming videos to classroom platforms, their lives unfold in front of screens. The internet offers immense opportunities for learning, creativity, and socialization, but it also exposes them to risks they cannot fully comprehend.

Children are especially vulnerable because they lack the cognitive maturity to assess digital threats. Persuasive design, targeted advertising, and online strangers posing as peers can easily manipulate them. Many apps marketed as "educational" or "free" quietly collect user data, building long-term behavioral profiles that may follow children into adulthood.

One of the most significant challenges is that online safety often depends on invisible choices made by adults — parents, teachers, developers, and policymakers. Parents or teachers expose children's data when they neglect privacy settings or fail to secure devices. When educational tools rely on third-party platforms, commercial databases may store personal information.

The psychological impact of early digital exposure is also significant. Constant online validation can affect self-esteem and attention. Social comparison begins earlier than ever, with children measuring their worth by likes, comments, and followers. Cyberbullying, once limited to schoolyards, now follows victims into their homes through smartphones and gaming chats.

Parental involvement is essential, but so is empowerment. Rather than banning technology, adults can teach children to navigate it with confidence. Conversations about consent, privacy, and respect should begin as soon as children can use a device. Digital literacy is the new safety lesson, as fundamental as learning to cross the street.

Every child deserves to explore the digital world safely. Protecting them means building systems that value innocence over influence and learning over profit.

Predators, Scams, and the Hidden Threats

The anonymity of the internet creates both freedom and danger. For predators, it provides a veil behind which manipulation can thrive. Online exploitation takes many forms, such as grooming, trafficking, fraud, and coercion, all made easier by instant communication and unregulated spaces.

Grooming often begins innocently. A stranger befriends a child through a gaming platform or social media site, gaining trust over weeks or months. Gradually, conversations shift toward secrecy, emotional dependence, or inappropriate content. A child may already have lost control by the time they realize what is happening.

Adults, too, face digital exploitation. Scammers use romance apps, phishing emails, and fake charities to manipulate emotions and extract money or personal information. Women and the elderly are frequent targets because of perceived emotional openness or lack of technical familiarity.

Human trafficking networks also exploit digital platforms to recruit, advertise, and control victims. Encrypted messaging apps and cryptocurrencies complicate detection. The dark web, an area of the internet accessible only through specialized browsers, remains a haven for illicit trade in images, identities, and personal data.

The fight against these hidden threats requires vigilance, education, and collaboration among governments, technology companies, and citizens. Law enforcement agencies are increasingly relying on digital forensics and AI to track online predators; however, prevention remains the most effective defense. Awareness training, parental monitoring tools, and open communication between children and guardians can expose manipulation before it escalates.

The internet's anonymity can hide cruelty, but it can also enable courage. When victims speak, communities can unite to identify patterns, report crimes, and advocate for stronger protections. The goal is not to instill fear but to replace ignorance with informed caution.

When the System Fails

Despite global awareness of online exploitation, legal systems often lag behind technological change. Jurisdictional limits, slow investigations, and outdated laws make prosecution difficult. Crimes committed across borders or in encrypted environments pose significant challenges to traditional policing methods.

Victims of digital abuse frequently face skepticism or blame. Women reporting harassment or revenge porn may be told to "just log off." Parents of cyberbullied children may encounter institutions that minimize harm because it happened online. In such cases, victims experience a second form of trauma: institutional neglect.

Tech companies, too, have struggled to strike a balance between user freedom and safety. Content moderation at scale is imperfect, as algorithms often miss harmful material or remove innocent content. Company policies often prioritize involvement over safety, resulting in responses to problems after they occur.

However, progress is possible. Some nations now mandate transparency reports from social media companies. Others have established specialized cybercrime units and victim hotlines. Nonprofit organizations provide digital shelters, online and offline spaces where survivors can seek help without fear of judgment.

For the system to succeed, three principles are essential: **accountability**, **accessibility**, and **empathy**. Accountability ensures that harm has consequences. Accessibility guarantees that every victim, regardless of background, can seek help. Empathy restores humanity to processes that too often feel automated.

Justice in the digital age is not only about punishing offenders; it is about rebuilding trust between individuals, institutions, and technology itself.

Tools for Protection

Technology, when used responsibly, can be both a shield and a sword. A growing ecosystem of digital tools helps protect vulnerable users from online abuse, exploitation, and manipulation.

For Women:

- Use **privacy-enhancing apps** and browsers that block tracking and hide personal information.
- Enable **two-factor authentication** and avoid sharing real-time location on social media.
- Keep personal and professional profiles separate when possible.
- Document harassment digitally and report it to both platform administrators and law enforcement.

For Parents and Children:

- Install **parental control software** that limits screen time and blocks explicit content, while maintaining transparency with children about why these measures exist.

- Teach children to question online interactions and never to share private information, photos, or passwords.
- Encourage open communication so that a child who feels threatened knows where to turn for help.

For Communities and Educators:

- Offer workshops on digital citizenship, empathy, and online safety.
- Advocate for privacy-focused school software that minimizes student data collection.
- Support peers who experience harassment or exploitation by listening and guiding them toward professional help.

For Everyone:

- Use strong, unique passwords.
- Regularly review app permissions and privacy settings.
- Learn to identify phishing attempts, deepfakes, and scam messages.

Protection begins with awareness, but it flourishes through community. When individuals share knowledge, the vulnerable become less isolated, and predators lose their power of secrecy.

Empowerment Through Awareness

The internet can be both a sanctuary and a battlefield. For women, children, and other vulnerable groups, it holds immense potential for empowerment, including access to education, economic independence, creative expression, and global solidarity. But safety must accompany opportunity.

Empowerment in the digital age begins with information. Awareness transforms fear into action. When users understand how systems work, they can challenge unfairness, report abuse, and make safer choices. When communities value empathy as much as efficiency, technology becomes a tool of liberation rather than oppression.

The future of digital safety depends on collective responsibility. Parents must educate, schools must protect, governments must legislate, and companies must design ethically. Above all, individuals must care enough to act — to look out for one another, to speak up, to teach, and to learn.

Women's voices and children's futures are too vital to be drowned in the noise of exploitation. By combining compassion with knowledge, society can create a digital environment where vulnerability is not a weakness but a call to protection.

To protect the vulnerable is to affirm our shared humanity. Every safeguard, every lesson, every boundary drawn online is an act of empowerment. It's a reminder that technology should serve people, not endanger them.

Ultimately, safety is not the absence of risk, but the presence of care. When awareness becomes culture, the digital world can finally reflect the dignity, equality, and compassion that define the best of humanity.

CHAPTER 12

THE FUTURE OF PRIVACY IN AN AI WORLD

Throughout history, every major technological revolution has required a corresponding moral awakening. The printing press demanded literacy. The industrial age demanded labor rights. Today's digital age demands responsibility.

We live in an era defined not only by tools alone but also by the ways we use them. Digital responsibility begins with awareness. The understanding that each click, share, and decision contributes to the kind of digital world we collectively create. Yet awareness is only the beginning. A truly safe and ethical internet requires a cultural shift: from individual caution to shared accountability.

Habits shape culture, and habits begin with people. Every time a user questions a source, protects their privacy, or refuses to spread misinformation, they strengthen the fabric of digital integrity. But culture expands when institutions, educators, corporations, and governments align around the same values of respect, transparency, and empathy.

Building a culture of digital responsibility means treating technology not as a force beyond our control, but as an extension of our collective humanity. It is about asking more profound questions: How can we design systems that reward truth over engagement, safety over speed, and empathy over profit?

The digital age offers unparalleled freedom, and with freedom comes duty. We must each become custodians of our shared digital environment, aware that every action contributes either to its pollution or to its preservation.

The Ethics of Everyday Technology

Ethics is often associated with grand decisions, such as corporate policies, government regulations, or public debates. Yet in the digital era, ethics begins with ordinary behavior. Every user wields influence. Every device holds power. The simple act of how we use technology — and how we treat others through it — shapes the moral landscape of the online world.

Defending the truth happens when a person verifies a fact. Protecting dignity happens when they pause before posting an insult. They are advocating for human rights by disabling invasive tracking features or supporting privacy-focused platforms. These seemingly small choices are ethical acts that have a ripple effect.

Ethical awareness must also extend to design and innovation. Developers, engineers, and entrepreneurs have immense influence over how billions of people interact with technology. Algorithms, user interfaces, and digital incentives shape human behavior. The choice to prioritize engagement over well-being has social consequences as significant as any law. A culture of responsibility demands that ethics become as integral to technology as code itself.

Even convenience has moral weight. The decision to allow a device to listen, a platform to track, or an app to store personal information involves consent. Understanding these trade-offs enables passive users to become active participants, not just consumers of technology, but citizens of the digital world.

Digital ethics is not abstract philosophy; it is everyday practice. It begins in the home, extends to the workplace, and echoes across society. Responsibility grows stronger each time awareness translates into mindful action.

The Role of Education and Leadership

If culture begins with awareness, then education is its foundation. Teaching digital literacy is no longer optional; it is as essential as reading and writing. From classrooms to boardrooms, individuals must learn not only how to use technology but how to question it; how to understand its biases, protect their data, and evaluate its effects on society.

In schools, digital education should go beyond technical skills. Children and teenagers need guidance in emotional intelligence, empathy, and online ethics. They must learn how to navigate identity, privacy, and peer pressure in a world where mistakes can have permanent consequences. Lessons about digital footprints, media manipulation, and consent should be as standard as lessons in history or mathematics.

Teachers, parents, and mentors play a vital role. They must model digital behavior that reflects integrity and care. When adults demonstrate how to balance online life with real-world relationships, young people learn that technology should enhance, not replace, human connection.

Leadership in the workplace is equally crucial. Executives and managers must establish ethical standards for data handling, communication, and transparency. A responsible company culture begins when leaders treat digital privacy not as a legal checkbox but as a moral responsibility.

Education and leadership together can create a generation that not only adapts to technology but also shapes it. The goal is to raise not just skilled users, but thoughtful citizens of the digital world—people who view technology to service, not a means of self-importance.

Corporate and Government Accountability

While individual awareness and education are powerful, lasting change requires institutional commitment. Technology companies and governments wield enormous influence over the rules of the digital world. They determine which behavior to reward, which rights to protect, and which values to prioritize.

Corporate responsibility begins with transparency. Companies need to disclose how they collect data, how algorithms function, and how their decisions impact users. Ethical design means building systems that protect privacy by default, not as an afterthought. It means limiting surveillance, curbing addictive design patterns, and giving users real control over their digital identities.

Governments also play a critical role. Strong privacy laws, data protection standards, and digital rights frameworks ensure that corporate power does not leave individuals defenseless. Policies such as the General Data Protection Regulation (GDPR) in Europe and emerging digital bills of rights worldwide are steps in the right direction. Yet regulation must grow as quickly as innovation.

Accountability should never be seen as opposition to progress. Instead, it ensures that progress benefits all. A society that values responsibility does not hinder innovation — it refines it. When corporations and governments act ethically, they create trust, and trust is the cornerstone of digital prosperity.

True digital accountability is not about control; it is about stewardship. Technology should serve humanity, not the other way around. The institutions that shape our digital world must remember that progress without principles leads not to freedom, but to dependency.

Bridging Generations and Divides

The digital revolution has created a generational gap unlike any before it. Older adults often feel overwhelmed by the pace of technology, while younger generations have been immersed in it since birth. Bridging this divide is crucial to fostering a culture of responsibility that encompasses everyone.

For older generations, digital inclusion is both a right and a necessity. Access to online banking, healthcare portals, and social media networks has become an integral part of everyday life. Yet without guidance, many seniors remain vulnerable to scams, misinformation, and privacy violations. Educational outreach programs, community workshops, and intergenerational mentoring can transform fear into confidence.

For younger generations, the challenge is unique. They are fluent in technology but not always in discernment. They may understand how to navigate apps and platforms, yet lack awareness of the long-term consequences — data permanence, algorithmic manipulation, and emotional burnout. Parents, educators, and mentors must provide balance, helping them use technology without becoming addicted to it.

Bridging the divide also means addressing inequalities across cultures, regions, and socioeconomic classes. Many still lack affordable access to reliable internet or digital education. Without inclusion, digital progress remains incomplete.

Technology should unite generations, not separate them. When knowledge flows in both directions — the young sharing innovation, the old sharing wisdom — the digital world becomes not just faster, but wiser. A culture of responsibility thrives on shared understanding across all ages and experiences.

Global Digital Citizenship

The internet knows no borders, and neither should digital ethics. Building a culture of digital responsibility means fostering *global citizenship*, an awareness that our online actions have a lasting impact on people everywhere.

Every post that spreads misinformation, every exploitative design, every data breach crosses continents. Conversely, every act of digital kindness, every verified fact, every responsible innovation, every respectful exchange, strengthens a shared global fabric.

Digital citizenship involves more than following rules; it is about embodying respect, empathy, and accountability in global networks. It means recognizing that cultural differences shape how people interpret privacy, expression, and technology — and that dialogue must replace dominance.

International cooperation is essential. Governments, tech companies, and civil organizations must collaborate on common standards for privacy, data ethics, and artificial intelligence. Initiatives like UNESCO's digital literacy programs and the United Nations' global principles on AI ethics represent early steps toward a unified digital moral code.

Global citizenship also means empowering those at the margins and amplifying the voices of developing regions, indigenous communities, and underrepresented groups. When everyone takes part, the internet fulfills its original promise: a space not just for connection, but for equality.

In a connected world, nothing isolates anyone's safety or dignity. Responsibility in one country can protect users in another. The digital age has made us neighbors in a shared virtual city, and citizenship comes with care, not just convenience.

The Road Ahead

As this journey through digital safety comes to a close, the most important lesson is that machines do not write the future. People write it — by the choices we make and the values we defend.

Technology will continue to develop faster than laws, institutions, or even language. Artificial intelligence will continue to grow more powerful, data will become increasingly abundant, and the boundaries between physical and digital life will continue to blur further. Yet the principles that guide humanity remain the same: honesty, empathy, curiosity, and courage.

Building a culture of digital responsibility means carrying these principles onto every screen we touch and onto every platform we shape. It means moving from passive awareness to active participation and understanding the risks of transforming systems.

Every person who speaks out against misinformation strengthens the truth. Strengthening the next generation requires each parent to teach their child about privacy. Ethical designers, fair policymakers, and compassionate citizens build a safer digital civilization.

The future of technology is not about control; it is about character. Machines will reflect what we teach them. If we embed ethics in algorithms and empathy in design, the world they create will mirror our highest potential.

The road ahead is uncertain, but it is also full of possibilities. Digital responsibility is not a final destination but a lifelong practice. It unites awareness, empathy, and action.

We should choose to build a world that respects privacy, values truth, and in which technology amplifies humanity rather than replaces it. The culture we build today will define the generations to come.

And in that choice — in every mindful action we take — lies the future of digital safety, and of the human story itself.

If this book has encouraged you to pause, even briefly, before clicking, sharing or agreeing, then it has achieved what I hoped it would.

About the Author

Lucy Lipari, M.A., is an educator, researcher, and author of six books spanning technology, history, fiction, and social impact. She holds a master's degree from New York University and has dedicated her career to helping readers understand the rapidly changing world around them. Her work explores the intersection of technology, humanity, and personal empowerment.

From the Author

Thank you for reading *Digital Safety in the Age of AI*.
The author wrote this book to give you clear, practical knowledge to help you feel confident and secure in an increasingly automated world. AI is evolving rapidly, and we are all learning together. I hope that this book empowers you to take control of your digital life—while understanding how technology shapes the world we live in.

If this book helped, inspired, or gave you a new perspective, your feedback is welcome and will help shape future work.

A Note on Developing Technology

Artificial intelligence and digital privacy tools are changing every day. Although we tried to provide accurate, research-based information at the time of publication, new updates, features, and risks will continue to appear.

This book only intends to provide general information, and you should not rely on it for professional, technical, or legal advice. Always verify settings, features, and privacy updates with your device or platform provider.

Acknowledgments

My heartfelt thanks to my husband, Richard, for his unwavering support and insight; to my readers who continue to inspire each new project; and to the writing community that has encouraged my work from the very beginning.

Books by Lucy Lipari and Richard Lipari

- *The Genghis Code*
- *North to Eden*
- *Dancing Words*
- *Eye of the Sun*
- *Killing Jack the Ripper*

Further Reading and Resources

To continue learning about digital privacy and AI, explore:

- Electronic Frontier Foundation (EFF)
- Mozilla Foundation – Internet Health Reports
- Consumer Reports Digital Lab

- Staying Safe Online – U.S. government cyber resources
- Research on AI ethics, digital well-being, and data transparency

COMING SOON IN THIS SERIES

Digital Safety for Teens: Protecting Your Privacy in the Age of AI

Teenagers today are growing up in a world shaped by artificial intelligence, constant connectivity, and social platforms that follow every move. While technology offers exciting opportunities, it also creates additional risks—many of them invisible.

Digital Safety for Teens is a practical, relatable guide that helps young people understand how their devices, apps, and online behavior affect their privacy, mental well-being, and future opportunities.

In this upcoming book, you'll discover:

- How AI tracks your habits, preferences, and routines
- Why "free apps" aren't really free—and what they take

- The hidden consequences of oversharing on social media
- How algorithms shape what you see, think, and buy
- Simple steps to protect your identity, reputation, and digital footprint
- Actual stories of teens who faced online challenges—and how they overcame them

Your digital life is part of your real life.
This book will help you protect both.

Request for Reviews

If you enjoyed this book, please consider leaving a review online. Reviews help other readers discover the book and support independent authors in creating future works. Even a brief review makes a meaningful difference.

Disclaimer

This book is based on research and information at the time of writing. Technology changes frequently. This book does not offer legal, financial, or professional advice. Readers should use their own discretion and consult qualified professionals when needed.

Made in the USA
Middletown, DE
04 November 2026

36543437R00056